Gas

Gas

Ian V Hogg

BB

Editor-in-Chief: Barrie Pitt
Editor: David Mason
Designer: David Allen
Picture Research: Carina Dvorak

Photographs for this book were especially selected from the following archives: page 2-3 Imperial War Museum, 6-7 US National Archives, 8-9 IWM, 10-11 Radio Times Hulton Picture Library, 12-13 IWM, 15 USNA, 21-22 RTH, 26-27 Library of Congress, 28-32 RTH, 33 Popperfoto Ltd, 35 IWM, 35 RTH, 36-39 40-41 Popperfoto, 40-41 USNA, 43 IWM, 44-45 IWM, 46-47 USNA, 48-59 IWM, 61 USNA, 62 IWM, 62 Popperfoto, 63-65 USNA, 66-67 RTH, 68 USNA, 69 IWM, 70 USNA, 71-79 IWM, 80-81 Popperfoto, 82-83 USNA, 84 Popperfoto, 86-87 USNA, 88-93 IWM, 94-97 USNA, 98-99 RTH, 100-111 USNA, 112-113 IWM, 114-115 USNA, 116-117 IWM, 118-119 Popperfoto, 121 RTH, 121-122 USNA, 122-123 IWM, 123 USNA, 124-125 IWM, 126 Artillerie Schule, 126-127 USNA, 128-129 IWM, 130-135 USNA, 136 American Chemical Society, 138-139 RTH, 140 USNA, 141 Popperfoto, 142 J. Topham, 143-151 RTH, 152-154 US Army, 156-157 Lib. of Congress, 158-159 Popperfoto

SBN 345-24392-7-200

First Printing: June, 1975

Printed in the United States of America

BALLANTINE BOOKS
A Division of Random House, Inc.
201 East 50th Street, New York, N.Y. 10022
Simultaneously published by
Ballantine Books, Ltd., Toronto, Canada

Contents

The appalling weapon

Introduction by Barrie Pitt

It was the late Captain Sir Basil Liddell Hart who first pointed out that the quite understandable public reaction against the use of gas as a weapon after the First World War may, in fact, have done a considerable disservice to those who fought in the Second. Certainly memories of the claustrophobic horrors inside close-fitting gas-masks in 1939 added to the horrific descriptions of the effects of mustard gas and Lewisite given us by eloquent and serious-minded lecturers, made one profoundly thankful in the years which followed immediately afterwards that at least our opponents were sufficiently civilised not to inflict that particular horror upon us.

But was such thankfulness justified? Was that apparent civilisation nothing more, in fact, than the result of correctly analysing the statistical tables?

This illuminating - and, incredibly enough, thoroughly entertaining - book raises some very valid points. I will not attempt to pre-empt any of Ian Hogg's arguments, but the comparisons of casualty figures caused by different methods of waging war given here will cause some surprise, and comparisons of recovery figures after incapacitation even more. Of course, as Ian Hogg says, 'the memory of pictures of blinded soldiers groping their way to the field dressing stations. ..and of course those individuals one met between the wars who suddenly went into paroxysms of coughing, smiled apologetically, tapped their chests and said "Gas, you know",' had an inevitable effect on public opinion, and gas warfare became the object of a universal loathing. Possibly this was because more people met still-living gas casualties than ever saw, for instance, the 'basket men' in some of the war hospitals who had lost all their limbs through artillery fire and spent the rest of their lives hanging from hooks on a wall.

So prose flowed about the horrors of asphyxiation and the readers blanched with shock, but "The remarkable thing is that nobody seems to have written a book about the horrors of say, the .45 bullet and the 9.2 shell. It was, apparently, quite acceptable, even admirable, to have the back of your head taken out by a lump of lead or to be disembowelled by a flying shard of steel, but entirely reprehensible to be made to sneeze or vomit."

Sneezing or vomiting is, of course, something of an over-simplification of the effects of First War gassing, for there was neither the quantity control to ensure that the gas was concentrated enough to be effective at all, or quality control to ensure that when it *was* effective, it was only sufficiently so to be temporarily incapacitating and not permanently so. And because of the popular reaction against any form of gas or chemical warfare, no research to that end was carried out during the inter-war years - although there was some in more lethal gases.

This was surely a pity - for there are gases which will rob those who inhale enough of it of all aggressive instincts, and one of course which will induce feelings of merriment (although the picture of enemy battalions rolling on the ground helpless with rollicking laughter is, alas, unfounded) and if the British, who the reader will be astonished to discover were by far the most efficient users of the gas weapon once they had caught up with the Germans, had devoted more time and research to the development and usage of such gases between the wars than they did to, say, the care and stabling of horses for their cavalry, they might have been in possession of a weapon as effective as the incendiary bomb and which would have left none of the post-war rancour resulting from "area bombing".

But this, perhaps, is romanticising in the opposite direction. The doctrine that the only way to win wars is by physical and permanent destruction of as many enemies as possible is far too entrenched in the military and political mind for such humane methods to be believed. The users - or perhaps even those who just suggest such a scheme - would lay themselves open to the charge of effeminacy at least and probably of decadence.

And perhaps those who set out to dominate the world prefer to deal with as small a population as possible anyway.

Explanation

The Redan battery at Sebastopol. Plans to attack the Russians in the Crimea with smoke were turned down by the Government

In 190 BC the city of Ambracia, in Epirus, was besieged by the Roman Army. Ambracia was strongly fortified, and after numerous attempts to breach the walls with battering rams, the Romans took to mining, digging a tunnel to the wall of the town and then extending a gallery beneath the foundations of a long section. Within this gallery the wall above was shored by timbers, and the intention was that when a sufficient length had been undermined, the chamber would be packed with combustible material which would then be fired. The supporting timbers would burn away and, devoid of support, a section of the wall would collapse, whereupon the besiegers would rush in and capture the town.

The Ambracians had other ideas; they had seen the excavated earth being dumped and realised what was afoot. They discovered the location of the Roman tunnel by careful listening, and, digging a counter-mine, broke into it. A large stone corn jar filled with feathers had been prepared, with a tube in its base connected to a bellows. After breaking into the tunnel, the Ambracians threw burning charcoal into the jar, inserted the mouth into the Roman mine, and pumped the fumes of burning feathers into the underground workings. The Romans were driven from the mine and the fortress was saved from that particular threat.

And so, 2164 years ago, we have the beginnings of chemical warfare, the

use of noxious fumes or gases to incommode, if not kill, an enemy. One conclusion which springs to mind is that in those unenlightened days there was no long-drawn-out argument about unfairness, inhumanity, frightfulness or any other liberal instincts; the enemy were at the gate -- indeed, they were under it -- and whatever came to hand was a legitimate weapon. In later years this pristine simplicity was somewhat blurred; during the Crimean War, for example, Lord Dundonald put up a proposal to smoke the Russians out of Sebastopol with sulphur fumes. The Government considered his proposal 'quite feasible, but the effects would be so horrible that no honourable combatant could use the device' and turned the idea down. There is an unconfirmed assertion that the Russians did attempt to use a substance known as cacodyl chloride, but in what way is not known, nor indeed does it appear to have had any effect if it was used, since none of the standard histories mention it. At much the same time, during the American Civil War, there was a proposal to use chlorine gas inside artillery shells, but this too was turned down as being too horrible to contemplate.

Chinese pirates operating in the area of Hong Kong in the late nineteenth century were in the habit of using 'stink-pots', small jars of burning sulphur and pitch which produced choking fumes, which they threw on to the decks of ships during an attack in the hope of putting the crew out of action. Similar stink bombs were occasionally used during the American Civil War, and they made a brief re-appearance in the Russo-Japanese War when the Russians attempted to smoke the Japanese out of mines they were attempting to drive beneath the defences of Port Arthur, a direct throwback to the Ambracians and their burning feathers.

The first attempt to put chemicals to use in an organised fashion was by the French police in about 1909; they were troubled in those days by a variety of 'anarchists' and other fanatics who tended to be difficult to extract from their hiding places without considerable shooting, and it seems to have occurred to some police official that a small grenade filled with tear-provoking chemicals would be useful to throw into such a retreat and incapacitate the criminal so that the forces of order could go in and get him. These had their first public outing in 1912 when a gang of criminals -- the notorious 'Motor Bandits' or Bonnot gang -- were cornered in a house in Choisy. After a brisk fusillade of gunfire with gendarmerie, a few well-placed grenades reduced the

gang to tears, and the police were able to capture them relatively easily.

This idea was taken up by the French army after seeing some specimens of the police grenade, and hand and rifle patterns of grenade were provided with the purpose of using them when dealing with fortifications. By firing or throwing these through loopholes in the work, the defenders would be rendered incapable of firing against an assault; this, at least, was the theory. An interesting point about this whole business is that neither the police nor the army appear to have taken much heed of producing any sort of mask or protective device; presumably, as far as the army were concerned, the occupants who were gassed would simply be by-passed by the attack, perhaps with one or two men detailed to keep them occupied by throwing more grenades in from time to time. The police, for their part, were probably relying on sheer numbers to be able to burst in, overpower the criminals, and burst out again before they were overcome.

Fume masks in use by a German naval gun crew

Which brings us, very rapidly to be sure, to 1914 and the First World War, in which gas was to become common-place. But before going further and becoming involved in the story, it might be as well to clear the decks, as it were, and get some basic definitions under our belts.

In the first place, although we speak glibly of 'gas', precious few of the substances used under this name were actual gases; and even those which took effect as gas were transported and handled in the liquid state. Some agents were solid, some were liquid, and some stayed in that form; not all gases were gaseous. However, we might as well go on using the word, as it has been hallowed by time and its connotations are understood by now.

The second thing to bear in mind is that it is not necessary to kill everybody in order to win a battle, by gas or by any other agency. In fact better results can be had by injuring men severely enough to prevent them fighting. An injured man inevitably attracts an uninjured one to his side to assist him, and (I regret to say) a fair proportion of these will be delighted for the excuse to assist their injured comrade, and themselves, out of the danger zone, thereby taking a respite if nothing worse. The injured man then demands the services of medical staff, ambulances, hospitals and so forth, and eventually is responsible for a good deal of national effort; produce sufficient injured men and you provide their country with a major logistic problem.

Another thing to bear in mind is that the gases which promise well in the laboratory don't always live up to that promise on the battlefield; goodness knows how many unfortunates have entered the next world through the door of a gas oven, but pumping town gas on to the battlefield would be a waste of time.

There are, broadly speaking, six groups of war gases; though in fact some of them, being versatile, can fit into more than one group. The first, taking them in order of their severity of effect, are the Lachrimators -- the

A gassed victim is carried from the battlefield

tear gases. Their effect is confined to the eyes -- producing copious weeping and thus a temporary blindness -- in general, but in some of them a high concentration can affect the respiratory system and end in being fatal. In the low concentrations usually found in an open air battlefield however, the tears flow but little more damage is done. Nonetheless, a man with his eyes streaming with tears is in no condition to offer a defence either of his position or himself, so that the lachrimators are useful war gases. Characteristically, they act very quickly -- in a matter of seconds from exposure to the gas the reaction begins -- and they can have an appreciable effect in quite dilute concentrations; 15 parts in ten million of air is quite sufficient to allow the presence of some lachrimators to be felt.

Next come the Lung Injurants or Asphyxiators. These act within the lungs, causing fluid to pass from the bloodstream and fill the lungs, preventing them from supplying oxygen to the blood and thus, in effect, drowning the victim. Moreover (as if that were not enough) some varieties are also systemic poisons which have an additional lethal value; a man gassed with a simple lung injurant may well recover if allowed to rest quietly in a clean atmosphere so that the fluids clear from his lungs; but the man gassed with a toxic injurant will be at greater risk since the poison will have been ingested into his system and will not be removed so easily.

Then come the pure Toxic Agents; those which have little direct effect on nose or eyes, but which penetrate the lungs, pass into the bloodstream, and from there spread through the body and paralyse the central nervous system. Most of the gases of this group are chemically related, being of the class known broadly as cyanides; and anyone who has read detective stories needs to be told no more about the effects of cyanide, either its virulence or its speed of operation. However, as we shall see, this is one of the classes in which performance in battle is not to be predicted from laboratory experiments.

Next come the Sternutators, or sneeze gases or 'respiratory irritants'. Most of these are not gases at all but solids, disseminated as extremely fine dust. They act very quickly; they act in minute concentrations; and they have an extremely pronounced effect, causing uncontrollable sneezing, nausea and vomiting. As with the lung injurants, some are simply irritants whose effect wears off when the victim is removed to a clean atmosphere, while others contain systemic poisons which have a delayed toxic effect. Most of this class are based on arsenic.

Fifth are the Vesicants, the 'blister gases', of which Mustard Gas is the most widely known and quoted. Here the effect is not solely through the lungs or eyes but places the whole body at risk. The effects of the vesicants, indeed, are usually multiple; some are lachrimatory as well as vesicant, some asphyxiatory, some are systemic poisons. Almost all are exceedingly toxic. The blistering effect is due to direct contact with the gas, from which it follows that the 'gas' in this case is, in fact, a liquid. When splashed about the battlefield it will land on men's skin and clothing, and after some hours blistering develops. In the eyes, a vesicant means rapid and permanent blindness. The vapour is less noticeable but has delayed effects in most cases. Until the arrival of the next group of gases, the vesicants were the bogeyman of the battlefield.

This final group is the 'Nerve Gases'; by strict definition, I suppose, they should be classed with the toxic agents, but the effects are so different that they are generally considered to be worth a separate grouping. Alone of the six groups they have never been tested in battle, having been developed during the late 1930s; and if the reports of the chemists are anything to go by, we should all be profoundly thankful. The nerve gases work, as

Result of a mustard gas attack

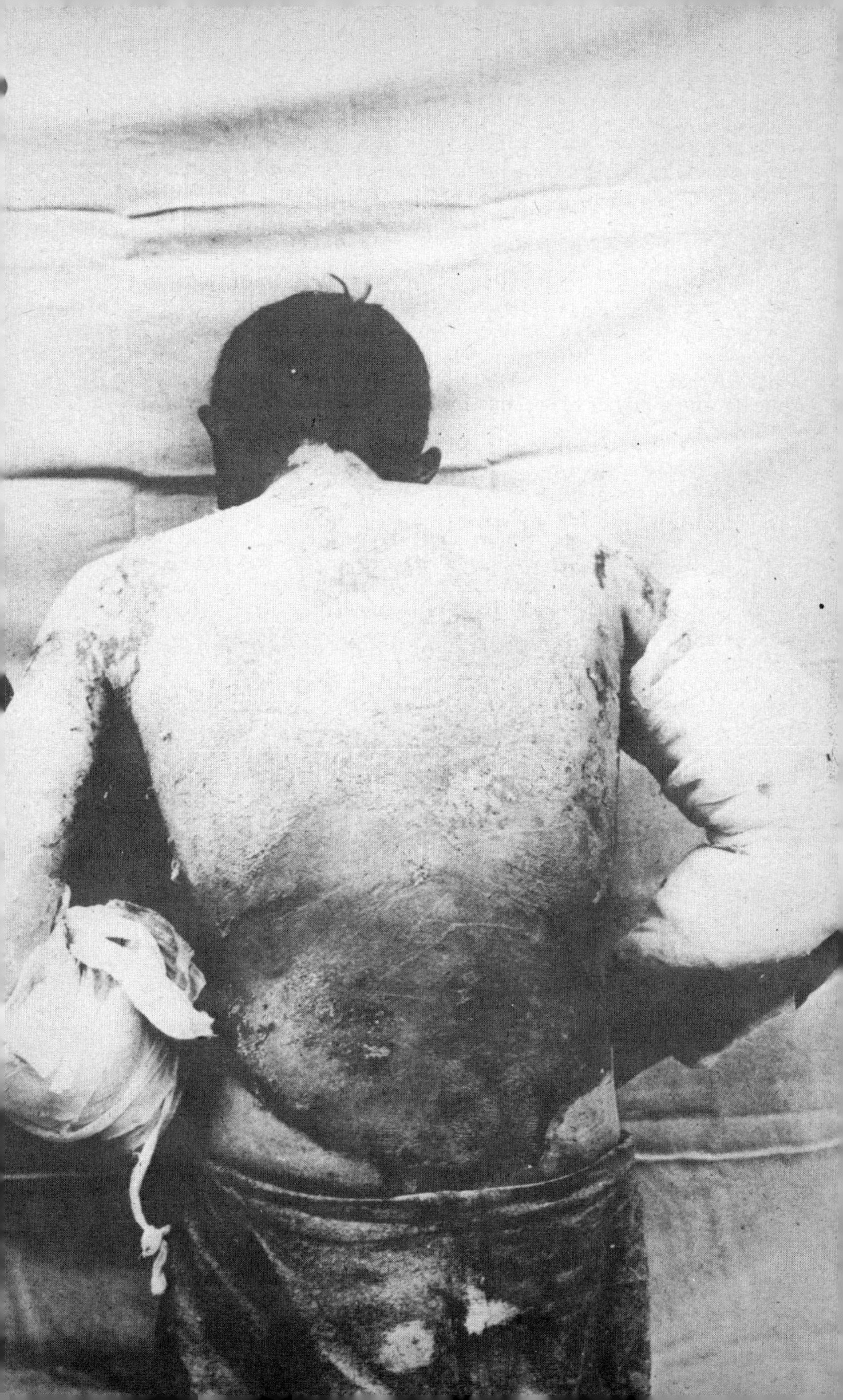

their name implies, by attacking the central nervous system. This system depends for its correct operation upon subtle chemical reactions, and the nerve gases upset the delicate chemical balance by destroying one of the vital compounds. This prevents the nervous system working as it should, and thus the body's functions are inhibited, leading to death in a matter of minutes.

The classification just given is a fairly exact one, and of use medically and chemically, but is of less use in war operations in the field. A commander contemplating the use of gas on an enemy is not particularly concerned as to whether he should employ a systemic poison or a lung injurant; his criteria are much simpler. Does it kill or merely incapacitate? Will my own troops be safe if they advance into the gassed area after the attack? That's all. As a result the military classification of war gases takes on a much simpler form. Gases are either Toxic or Irritant, Persistent or Non-Persistent. What sort of gas it is, what the chemical formula is, these are of no particular interest to the man using it, providing he knows it is, for example, a non-persistent irritant. On the other hand the man on the receiving end wants to be able to identify it more precisely so that he can take the proper steps to treat casualties.

The terms 'toxic' and 'irritant' speak for themselves; they either kill or incapacitate temporarily. Nerve gas is toxic, tear gas is irritant. But 'persistent' and 'non-persistent' are terms which define not its actions on the human body but its behaviour in the open air. Pour some methylated spirit on to a warm stove and it evaporates almost instantly -- it is non-persistent. Pour some engine oil onto the same stove and it will sizzle and smell for a long time before it entirely evaporates -- it is persistent. In a similar way, when war gases are released, either by shell or spray or bomb or any other means, they behave in one of these two ways. Either they soon disperse due to their lightness or due to warmth or rain or wind effects; or they soak into the ground, into buildings, clothing, wood, and continue to contaminate the area and endanger life for some considerable time. The dividing line between persistent and non-persistent gases is very much a matter of personal opinion, but as a general rule it can be said that if a gas remains in a sufficient amount to require protection to be taken against it at the end of ten minutes from its release, then it is a persistent gas. Some persistent gases can remain active for days, or even months, depending upon weather conditions. Indeed, mustard gas, after soaking into porous concrete protected from weathering, has been found dangerous after twenty years.

The military significance of this is important; if you intend to invade the gassed area and use a persistent gas, then your own troops need protection from the long-term effects of the gas. If, on the other hand, a non-persistent gas is used, then the attackers can dispense with protection while arriving on the scene rapidly enough to take advantage of the effects of the gas on the defenders.

It ought to be borne in mind, of course, that persistence or non-persistence has no bearing on the physiological effect of the gas; one can have toxic persistent or toxic non-persistent, irritant persistent or irritant non-persistent at will.

So the commander, instead of arguing about what precise chemical to use, simply decides on, let us say, an irritant non-persistent gas. And in order to keep it simple all the way down the line, gas munitions are all classified and marked according to this simple division. Thus in United States service, toxic gas munitions are marked in red, irritant gas in green; persistent gases are marked by two coloured stripes, non-persistent by one; all very simple and logical. Similar systems have obtained from time to time in other armies. One of the earliest was that developed by the Germans during the First World War and continued in a slightly modified form in the Second. Munitions -- for example, artillery shells -- containing gas were marked with a coloured cross; White Cross indicated 'harmless' gas -- in other words lachrimators. Blue Cross indicated 'slightly harmful' (German classification, not mine) by which they meant sternutators; Yellow Cross covered the 'moderately harmful' vesicant gases (though some people might dispute the tag 'moderately harmful' as applied to mustard gas) and finally Green Cross was the 'severely harmful' lung injurants and systemic poisons. A refinement was the splitting of the groups by numbers to indicate particular gases -- eg, Blue Cross One and Blue Cross Two -- but these were largely for the benefit of supply organisations and the artillerymen were simply concerned with the colour of the cross.

The system, while good, has one obvious defect; it makes no distinction between persistent and non-persistent gases. But in fact, due to the particular gases the Germans were using at that time, it so happened that classification was automatic; by chance, all the White, Green and Blue Cross agents were non-persistent, while the Yellow Cross was persistent. In fact one type of filling was re-classified from Yellow to Green solely on the grounds of its persistency.

This division into persistent and non-persistent carries with it one small defect for the attacker; it gives away his intentions to a large degree. If an area is attacked with a persistent gas it follows that the attacker is unlikely to move troops against it, and is merely out to create a gas barrier to, for example, protect his flank. Thus the defenders can be sure that wherever the enemy assault is going to fall, it will not be in the area saturated with persistent gas, and he can plan his dispositions accordingly.

As a last section of this short review of the technical aspects, we might just look at what goes to make an effective war gas. As already said, the fact that gas is lethal under laboratory conditions is no guarantee that it will be an effective gas in warfare. During the First World War it has been estimated that between 3,000 and 3,500 different chemical substances were investigated and evaluated for use as possible war gases, and of this vast array only thirty-eight compounds were selected for use, and of those only a dozen or so achieved any worthwhile results in the field.

The prime requirement is, of course, that the substance should be either toxic or have some pronounced physical effect which will incapacitate the victim. It should have this effect as rapidly as possible and in as low a concentration as possible. In addition, it should ideally have a multiple effect; it could, for example, be a powerful lachrymator and a lung injurant as well, so that protection against its effects becomes more involved and difficult. It should also be sufficiently persistent to stay at least long enough to take effect on its victims, and, as a corollary, should have as long-lasting an effect as possible so as to place as much strain on the victim's medical services as possible.

Assuming that all these things can be achieved, there are certain other characteristics which will improve the effect of the gas. It is advantageous, for example, if it has no characteristic or easily recognised smell which will warn its victims before a dangerous concentration has been built up. It helps too if the gas is invisible. And, of course, it should be cheap, easy to make, safe to transport and handle, and all the other things which the economics of warfare demand.

If all this can be achieved, then an effective weapon is to hand. But remember; twelve effectives, from thirty-eight possibles, from three and a half thousand substances; the road to effective weapons is never a simple and direct route, but more a species of maze, with blind alleys and time-consuming detours. The following pages attempt to chart some of the more interesting convolutions of the path.

Demonstration

On 20th December 1911 three men in a stolen automobile intercepted a bank messenger in a Paris street, shot him dead, stole his satchel while threatening bystanders with pistols, and departed the scene as rapidly as they had come. This was the debut of the 'Motor Bandits', and between then and May 1912 they killed another six people, wounded another dozen, and carried out a score of robberies, invariably escaping in a stolen car. Eventually two of the ringleaders, Bonnet ('The Demon Chauffeur') and Dubois, were trapped in a garage at Choisy-le-Roi on 28th April 1912; after a short and sharp siege they were dynamited into oblivion by the exasperated police. The leadership then fell on their erstwhile deputies, Garnier and Vallet, and these two together with others of the gang were trapped in a villa in Nogent-sur-Mer on 14th May and surrounded by over a thousand police and troops. In order to capture as many alive as possible the police produced a number of grenades containing tear gas and threw them into the building. Garnier and Vallet in fact died during the siege but the rest of the gang were taken, tried, and suffered a variety of penalties for their activities.

Thus it came about that the first use of gas in modern times was in what is now called 'Aid to the Civil Power'; and we might note in passing that probably as much gas has been used in that field of activity in the ensuing years as in the more direct forms of warfare.

The French grenades were soon adopted by the army with a view to using them against fortifications, as we have already said. The final version produced was a 26mm calibre rifle grenade mounted on a slender rod which could be slipped into the muzzle of the rifle and discharged by a special blank cartridge. These contained 19 cubic centimetres of Ethylbromoacetate, a substance widely used at that time in various chemical processes and one whose tear-provoking characteristics were well known to chemists.

To appreciate the French Army's adoption of this grenade, it must be recalled that in 1912 the fortress was still widely considered to be a vital form of obstacle, and the continental countries of Europe were liberally furnished with such defensive works on every frontier. The classic system of defeating these obstacles was to trench one's way up to the protecting ditch and then attempt to enter the fort by breaching the walls, scaling them, or entering by way of the various casemates and galleries fronting onto the ditch. The French intention was to fire these grenades through the loopholes of those galleries which covered the ditch, so as to prevent the occupants from offering any resistance while the final phase of the attack was being pressed home. It was very sound reasoning; anyone who has ever risked claustrophobia by exploring the subterranean galleries of these old-time forts will readily agree. In such confined and airless, gloomy and unventilated spaces a few milligrammes of lachrimatory gas would soon have taken effect. But, in the event, the French never had the opportunity to adopt this system of attacking forts, so we will never know how effective it might have been.

The German army were aware of these grenades, however, and when the First World War broke out in August 1914 they were on the watch for them. It seems likely that the French did try using them once or twice in open warfare conditions, but they had no effect and were apparently discarded as useless. But the Germans were alarmed by reports from the front of deaths of troops apparently due to asphyxiation after shelling, and they despatched a number of medical and chemical experts to check. These examiners discovered that the deaths were due to carbon monoxide, a product of the detonation of high explosive, entering dugouts and there overcoming the occupants, and that this was merely an unexpected by-product of the use of high explosives. At the same time, the Allied newspapers caused more alarm in Germany by printing highly coloured stories of a new explosive substance called 'Turpinite' which, it was claimed, killed by its fumes those whom it failed to kill by its detonation. Just what Turpinite was is not entirely clear, but Eugene Turpin was a prolific inventor of explosives (he had developed Melinite, the standard French high explosive for shell filling), and it seems that Turpinite was a liquid explosive, a variation of an earlier patent called 'Panclastite' which was known to give off dangerous vapours when used in mines or confined spaces. In spite of the rapturous tales in the press, Turpinite was only adopted as a shell filling by the French because there was insufficient of the more usual type of explosive available, and it was far from successful; but if it did nothing else it kept the German army worrying about gas effects.

Now one of the most remarkable features of the First World War was the way in which the demand for munitions took all the combatants by surprise. Even Germany, who were probably better prepared than anyone else, failed to realise what sort of demand modern war would place on the production of weapons and ammunition. This was made worse by their belief that they would walk all over France as they had done in 1870, only quicker, and that the campaign would be over in six or eight weeks. Because of this viewpoint, they neglected to mobilise industry for the volume production which later became necessary, and before 1914 was over the failure made itself felt, notably in the provision of artillery ammunition and explosives. In order to cope with the sudden and enormous demand for shells, it became necessary to manufacture them rapidly from inferior material, and numbers of cast iron shell were produced. These, owing to requirements of strength on firing, were thicker in wall section than the normal steel shells, and thus when charged with the usual types of explosive carried less than the desirable payload. When they detonated, the effect was thus less lethal owing to reduced blast and poor fragmentation. As a result the chemical industry were asked to investigate the problem and see if they could find some substance that could be filled into these cast iron shells to improve their effect and, at the same time, economise in high explosive.

One of the standard projectiles fired by the German artillery was the 105mm 'universal shell'. This was basically a shrapnel shell, designed to burst in the air and shower the enemy troops beneath with several hundred lead balls at high velocity. Such a shell was standard in every army, but the Germans had taken the idea a step further; most designs had the lead balls packed in resin, so as to stop them rattling about and also to provide a puff of smoke when the shell burst so as to assist the gunners in adjusting the range. The universal shell had the balls packed in powdered TNT; when used as a shrapnel shell, the TNT burned and gave a black smoke puff. But by adjusting the fuse, the shell could be made to operate on impact, when the TNT detonated and scattered the balls and fragments of shell in all directions and with great violence.

In September 1914 a Professor Nernst, who appears to have been employed as a chauffeur to the Director of Medical Services of the German First Army at the time, proposed replacing the TNT in a universal shell by an irritant powder, dianisidine chlorosulphonate. This would economise on TNT and would produce a lachrymatory effect when the shell was used as shrapnel. The idea was accepted, and Nernst was sent to a chemical factory at Leverkusen to superintend the production of the shells. A trial was carried out at Wahn, near Cologne, on 12th October and a supply were then sent to various batteries in the Neuve Chapelle area, and in an attack there on 27th October the shells were fired.

It seems that the idea was a complete failure. The Allies noticed nothing out of the ordinary, and, as a clincher to the whole affair, General Falkenhayn's son is said to have won a case of champagne by remaining in a cloud of the substance for five minutes without displaying any signs of discomfort. Since shrapnel shell, for best effect, should burst some thirty to sixty feet above the ground, there is every likelihood that the irritant drifted away before reaching the ground and thus had no effect at all. In any event, the idea was abandoned forthwith and never revived.

Investigation now began in attempts to discover something that could be put into cast iron shell, and the Kaiser Wilhelm Institute for Physical Chemistry and Electrochemistry in Dahlem, a suburb of Berlin, was asked to undertake a variety of experiments. Professor Emil Fischer suggested the use of cacodyl chloride, and during investigations into this substance a violent explosion occurred which killed Professor Sachur and blew off the hand of Professor Just. Work on this line of exploration stopped forthwith.

But while this had been going on a Doktor von Tappen of the institute had, through the agency of his brother who was Chief of Staff at GHQ, proposed putting xylyl bromide, a tear gas, into artillery shells. Tappen was familiar with this substance since he had prepared a thesis on it for his research degree, and he was confident that by placing a quantity inside a shell so that it would be distributed as vapour when the shell burst, it would have a considerable lachrimatory effect at the target. His original idea was to employ a large number of heavy-calibre mortars to fling the projectiles, but this was turned down and conventional artillery proposed instead.

Tappen now interested Professor Haber, Director of the Institute, in his ideas, and Haber became most enthusiastic; so much so that he eventually became the king-pin of German gas warfare. Haber attended a trial of the Tappen gas shells at the Artillery School Ranges at Kummersdorf in December 1914, and observed that to get the best effect it would be necessary to fire large numbers of shells simultaneously in order to produce a high concentration of the vapour at the target. One can well imagine the thoughts of the professionals at Kummersdorf being told how to conduct their shooting by a civilian professor; but in fact, Haber was perfectly correct in his deductions, and the combination of Tappen's mortar idea and Haber's concentration, in British hands, was to give the Germans a nasty fright in later years.

Since his idea of concentrated fire

An experiment in progress at the Kaiser Wilhelm Institute in Berlin, birthplace of German gas warfare

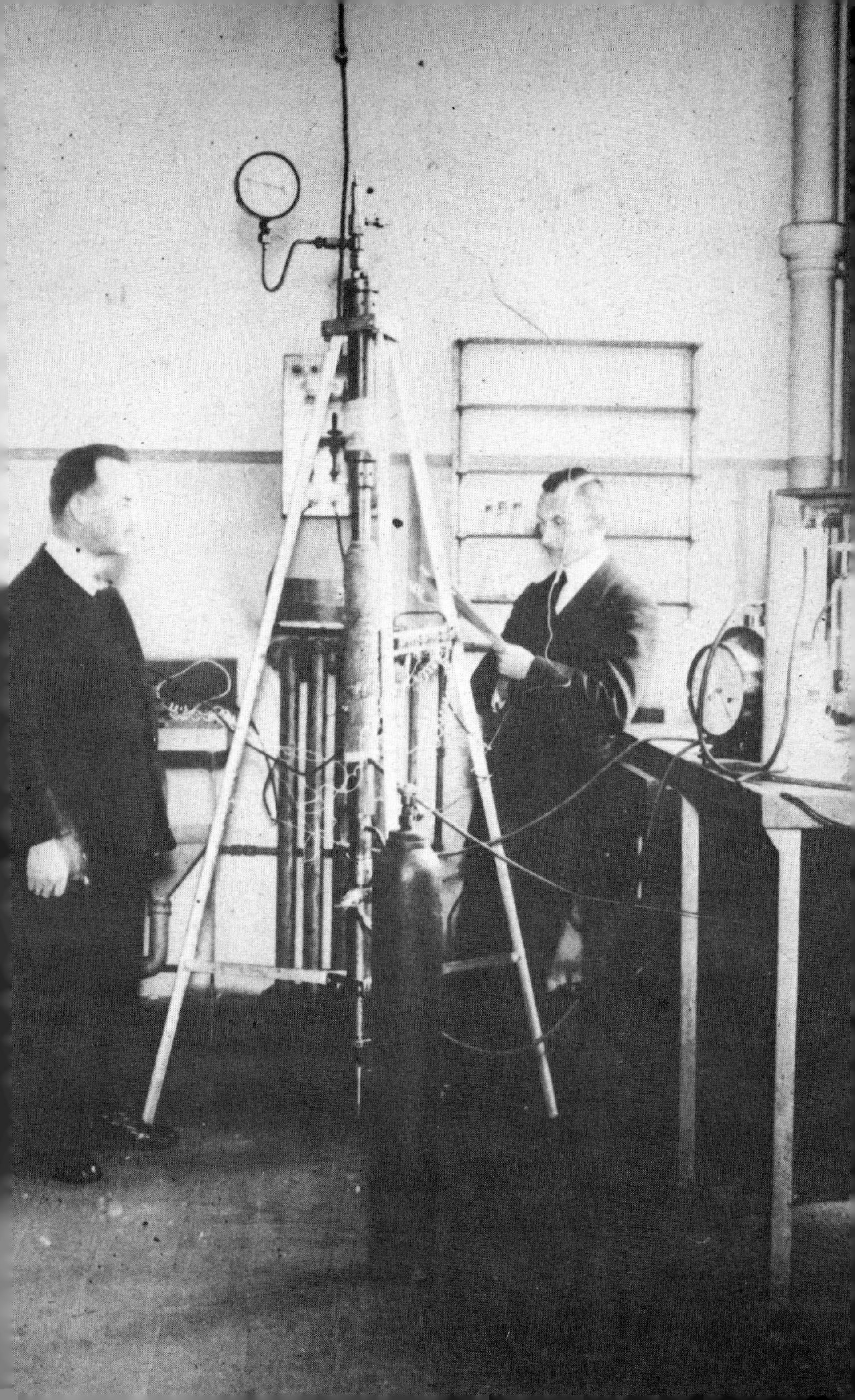

seemed to receive a stony reception, Haber next suggested simply putting a suitable gas into cylinders, taking it to the front, and opening the taps at the crucial moment. This, he claimed, ought to get a potent concentration onto a wide area. The High Command considered this, accepted the idea in principle, and then passed it back to Haber for him to develop it; reading between the lines of the reports, one has the feeling that they did not think much of that idea either, but gave Haber the go-ahead on the assumption that it would keep him quiet for the rest of the short war and stop him coming back with more silly ideas. Which is how soldiers felt about professors in 1914, and not only in Germany.

The gas shells passed their trials and were placed in production, to be used as opportunity offered. The design of the shell was governed by some conflicting factors; in the first place, the first Hague Conference of 1899 produced a resolution whereby the contracting powers agreed to abstain from the use of projectiles 'for the diffusion of asphyxiating or deleterous gases', and Germany was one of the contracting powers. But at the conference it had been pointed out that the detonation of any explosive produces some deleterous gas, and eventually it was agreed that if the splinter and explosive effect of the shell was greater than the gas effect, then it did not infringe this ruling and would be acceptable. As a result, the new shells had to have a good splinter and explosive effect, which meant a fairly heavy charge of explosive.

The other principal problem lay in the question of what would the shell do in flight? The artillery were fairly adamant that filling shells with liquid was a forlorn hope; the stuff would shift about in flight and so upset the ballistics that the shell's trajectory would be quite unpredictable. In this, they were quite right; liquid-filled shells will do funny things in the air, and the art of the ordnance engineer has to be called in to make it behave properly, and in the end this is what had to be done. But in the beginning it was thought to be somewhat risky, to say the least. Then, of course, the construction of the shell had to be such that the chemical was kept away from the explosive, or goodness knew what might happen; explosives and war gases are both substances tending to instability and nervousness, and contact between the two would be distinctly hazardous.

Professor Haber, Director of the Institute

Eventually the design was settled, and the 'Tappen Shell' or 'T-Shell' were ready for use. By an extension from this name, the question of what to tell the troops about what was inside the things was solved; the shell contained 'T-Stoff' and was stencilled with a large block 'T' to identify it. And this, in turn, led to the German practice of calling various gases 'Stoff' with an identifying letter.

The T-Shell was a standard 15cm howitzer shell which carried a lead cylinder filling most of the body. The cylinder contained 7 lbs of xylyl bromide, and above it, in the head of the shell, was a charge of 3.3 lbs of TNT as a burster. The lead cylinder was necessary in order to prevent contact between the chemical filling and the TNT, and it was secured in the shell by paraffin wax.

All that was wanted now was an opportunity to use the T-Shells, and here the problem was to persuade the soldiers to give it a try. There was a considerable difference of opinion on the subject; some considered it wrong on humanitarian grounds; some felt it was sailing too close to the Hague Convention for comfort; some just didn't think much of the idea; and others, more percipient, were a little worried about starting something which might well rebound to their disadvantage. Eventually it was agreed to give the T-Shells their chance on the Eastern Front.

During the winter of 1914/15 the Eastern Front had become as rigid as the Western, and Hindenburg planned an attack in the direction of Warsaw to get things moving once again. One attack would go in at Bolimov, on the Bzura river; if it succeeded, well and good. If it failed, it would at least have kept the Russians occupied while another attack in East Prussia got underway. And since the Russians at Bolimov had a good strong position, the T-Shell would be used to soften them up. 18,000 T-Shells for the 15cm howitzers were sent to Bolimov, and on 31st January 1915 the attack began.

It will serve no useful purpose here to follow the course of the battle; suffice it to say that a week after it began the German army was back where it had started, leaving 20,000 casualties behind. The T-Shell failed miserably. The basic reason is so simple that it seems surprising that nobody thought of it; the weather was so cold that the xylyl bromide failed to vapourise when the shells burst, and apart from one or two minor local effects in isolated places, the shells had no effect and no bearing on the battle whatever. No casualties were caused by them, and the only result was to alert the Russians to the possibility of gas attacks and cause them to begin work on providing some sort of protection for the future.

In spite of this initial set-back the T-Shell was not abandoned. The reason for its failure was appreciated and production continued at a low priority so as to build up more stocks for a fresh venture. It appears from some reports that it was used against the British at Nieuport in March 1915, but once again it seems to have gone unnoticed by the recipients. It is not without interest to see that neither the Russians nor the British made any announcement about the use of gas on either of these occasions.

While this had been going on Professor Haber had been bustling about on his project for discharging gas from cylinders. The only poisonous gas which was at all suitable and available from commercial production was chlorine, since it was widely used in industry and there was ample plant for its manufacture. A few cylinders were filled and an experimental discharge made at Hasselt, in Belgium, when Haber himself walked into the cloud and was slightly gassed. This seemed to satisfy everybody that the idea worked and, doubtless in the cause of secrecy, no further trials were made. This was a serious mistake, since it meant that the possible effects of a mass discharge were neither understood nor appreciated by the army; as a result they neglected to provide their troops with any form of protection and they neglected to make arrangements to take advantage of the surprise effect that the forthcoming attack would have.

Haber now rounded up all the cylinders he could commandeer, and trained a special pioneer company in their use. When the High Command finally decided on the location of the gas attack, Haber and his pioneers arrived with 5,730 cylinders containing 168 tons of chlorine, and, in great secrecy, installed them in the front line trenches. Pipes led from the cylinder valves across the trench parapet, and one pioneer was in charge of every ten cylinders. In all, the front of the attack was six kilometres long.

Haber's choice of chlorine, although governed by the question of availability, was a good one, since chlorine fulfilled many of the desirable features of an effective weapon. In the first place it was fairly lethal; non-persistent; immediately effective; and extremely volatile. Its only disadvantage was that being a highly active chemical it readily combined with other substances, and this makes it easy to counter. But since this was going to be the first attack of its kind, there would be no likelihood of the victims having any means to hand with which to counter it.

In spite of the secrecy with which

Haber had surrounded all the activity, it was obviously impossible to carry 5,730 large cylinders of gas into the front line without causing a lot of chatter, and inevitably some of the chatter crossed No Man's Land with prisoners. Sir Horace Smith-Dorrien, Commanding the British Second Army, noted in his diary that he had received reports from the French army, which had been collected both by their agents and from prisoners captured by them, that the German 26th Corps was about to attack. "The details given by the prisoners are so voluminous and exact that I am sure they are true," he wrote. "The prisoners described how they had arranged in their trenches batteries of enormous tubes of asphyxiating gas, and how, at a given signal, when the wind was in the right direction, all these tubes are to be opened.... " But no attack materialised on the promised day, 15th April, and the belief grew that it was some sort of hoax.

The nub of the matter was there, in the prisoners' statements: "When the wind was in the right direction." The attack at Ypres, between Bixschoote and Langemarck, was supposed to take place on 15th April, but the wind on that day was in the wrong direction and the attack was postponed from day to day. At last, on 22nd April, the wind changed in the afternoon and without further delay the attack was ordered. At half past five in the afternoon, the German pioneers grasped their valve keys and, running from cylinder to cylinder, opened the valves when the signal -- three coloured flares fired from an observation balloon -- was given. Within five minutes the cylinders were empty and 168 tons of chlorine were rolling towards the Allied lines.

Give the German army its due; it had an unerring instinct for seeing the weakest section of the line in front and going for that point. Here they had a weapon of unknown potential, of terrifying aspect and effect; and in front of them was the French 45th Algerian Division, made up of Zouaves and other native troops, steady enough under fire but when confronted with something which looked like witch-doctor's magic they broke and fled in screaming panic. The unlucky ones fell, clutching their throats, and died; the lucky ones kept on running until they outdistanced the cloud, cleared the battlefield and arrived in Ypres, six miles away, where their horrified stories lost nothing in the telling. The chlorine cloud had punched a five-mile hole in the front.

Technically the attack was a complete success; the official despatch of Sir John French reported: "The effect of these poisonous gases was so virulent as to render the whole of the line held by the French Division practically incapable of any action at all. It was at first impossible for anyone to realise what had happened. The smoke and fumes hid everything from sight, and hundreds of men were thrown into a comatose or dying condition, and within an hour the whole position had to be abandoned, together with about fifty guns. I wish particularly to repudiate any idea of attaching the least blame to the French Division for this unfortunate incident."

Tactically, however, the attack fell a long way short of what might have been achieved. Owing to the constant postponements, the German troops had 'gone off the boil'. They were physically tired from constant standing-to for an attack which was always put off; they had lost the fine-honed edge of trained perfection; they were psychologically undermined, and had lost confidence in the promises of the gas enthusiasts. The attack started so late in the day that darkness fell before the German troops could reach their objectives. And to make matters worse the German artillery had weighed in with some T-Shells at the wrong moment and had thus held up the advance. The staff, having little faith in the promised effectiveness of this new weapon, had made no provision for

The First Hague Conference of 1899 proposed an agreement for contracting powers to abstain from the use of projectiles for the diffusion of asphyxiating or deleterious gases

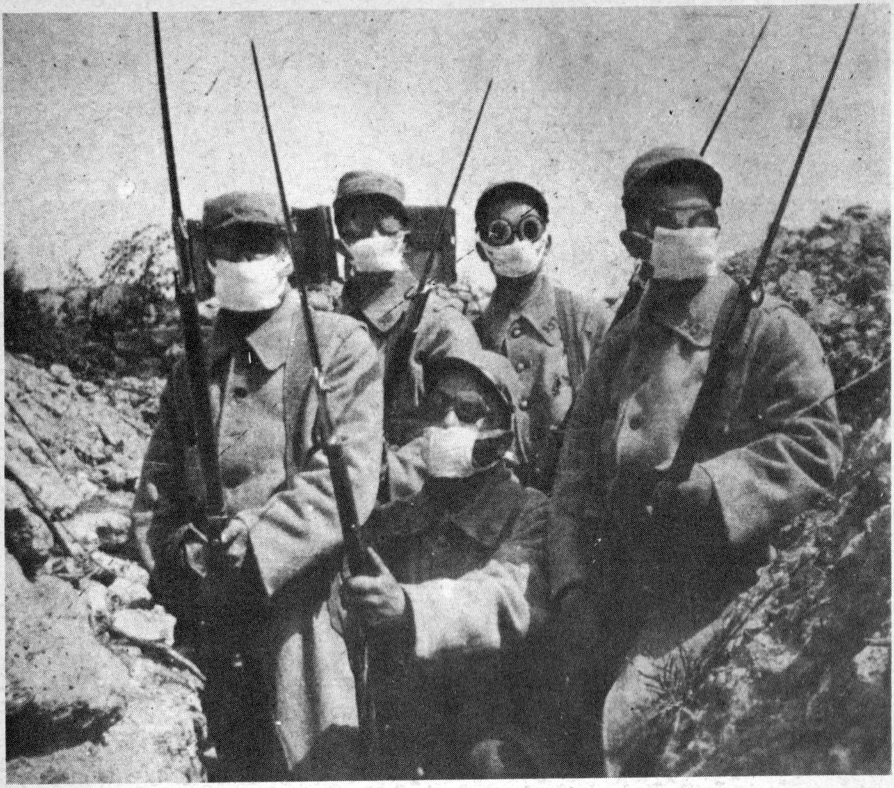

reserves to move up quickly to exploit success; and when the reserves did move it was in the dark and many of them lost their way. In the darkness the German troops had no idea that the Allied front had collapsed, and instead of pushing forward they contented themselves with reaching their stated initial objectives and digging in for the night. As a result, the Allies managed to rally enough men during the night to hold the line.

The next day saw more clouds of gas, but by this time it had been identified and the medical services had hurriedly improvised face-pads of surgical lint and tapes which could be dipped in water and tied across the nose and mouth to give some degree of protection. Consequently when the next large cloud fell on the Canadians on 23rd April, they were able to hold their positions in some order and repulse the German assaults.

Early French attempts to protect against gas; *Above:* in the trenches, *Right:* in mobile war

Eventually the Second Battle of Ypres came to an end; the line had been straightened, and 20,000 men gassed of whom it is estimated that 5,000 died. They died unpleasantly too, for chlorine, against unprotected men and in the concentration achieved by the cloud, killed by strangulation, with considerable pain. This aspect of the attack, the violent, painful and unprotected deaths, was the root cause of all the millions of words of propaganda and misunderstading which has conditioned thinking and clouded the issue ever since. Many writers since have condemned the Allied commanders for failing to secure some protection for their troops when they knew, from the prisoners' statements, that the Germans were preparing to use gas; but as will become apparent, you cannot protect against a gas unless you know with some degree of accuracy what sort of gas it is. Moreover, one must try to put oneself in the position of a soldier on 21st April 1915, when gas warfare was unheard of; how do you start to protect yourself against something entirely outside your imagination?

Needless to say, in the days that followed the old saw about 'All's fair in Love and War' was quickly forgotten as everyone rallied round to vilify the Germans for what they had done. The Hague Convention was freely brandished about, though the Germans were quick to point out that the phrase in question was 'the use of projectiles the object of which is the diffusion of asphyxiating or deleterous gases' and gas cylinders, by definition, were not projectiles, and therefore their action was not counter to the wording of the Convention. They might also have pointed out that when they *had* used projectiles -- unsuccessfully -- nobody had complained. In fact the protests against the German action were the last protests of the amateurs against the professionals. Now it was finally realised that when national survival was at stake there was no price which could not be paid, no limits of good behaviour, no concern for 'the rules'. The cloud gas attack at Ypres marks more than the start of large-scale gas warfare; it marks the end of the Gentleman's Wars and the beginning of the era of Total War, war in which the end justifies the means and in which it is akin to treason not to use any weapon which promises to achieve the desired result.

Experimentation

The first British protective masks

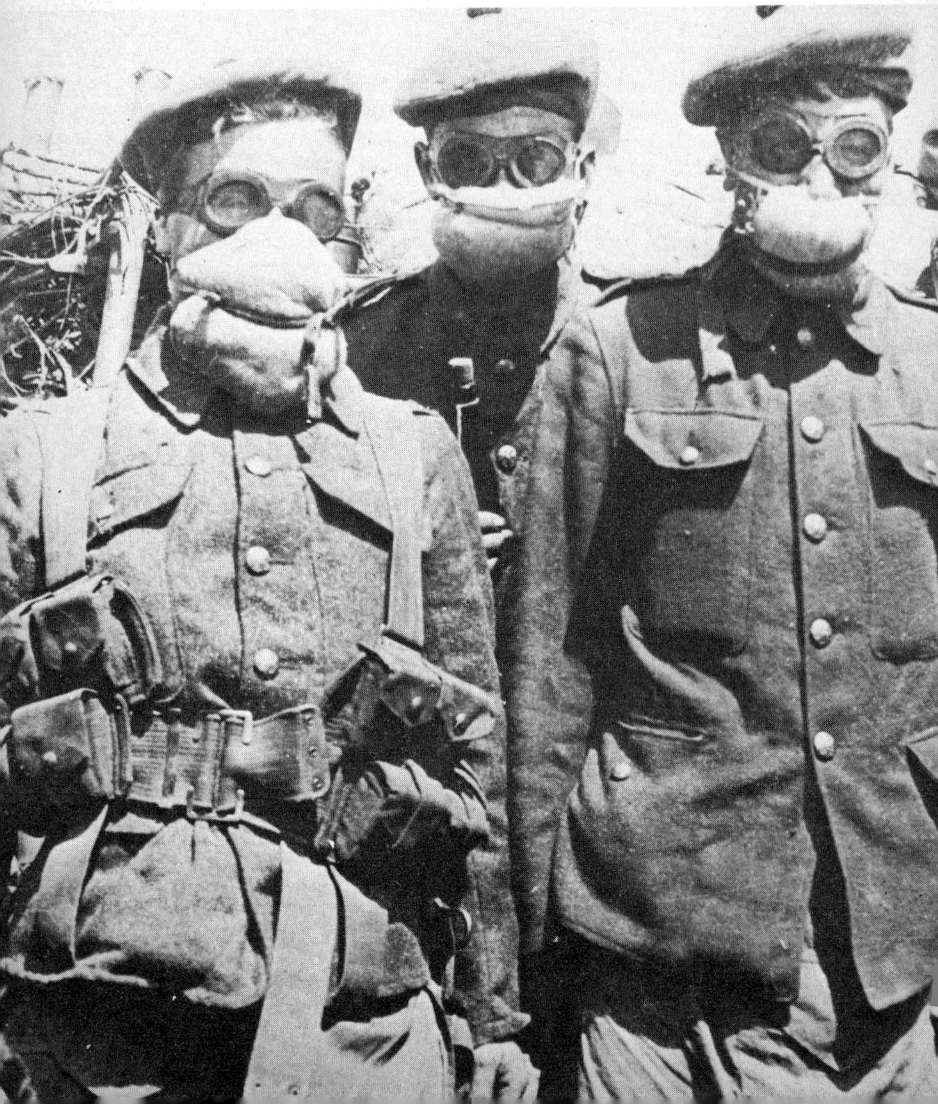

The reactions to the Ypres cloud attack were fairly predictable; the French and British immediately began production of a simple form of protective mask. Within a fortnight every British soldier had a cotton pad soaked in sodium carbonate and sodium thiosulphate which, tied over his nose and mouth, afforded protection against chlorine gas. This was a temporary expedient, and by 10th May the troops in the Ypres sector were issued with the 'Black Veil Respirator', a fourfold piece of black veiling a yard long and containing a cotton pad soaked in sodium carbonate, glycerine and water. This was wrapped around the face so that the pad covered the nose and mouth, and it gave reasonable protection against inhaling the gas.

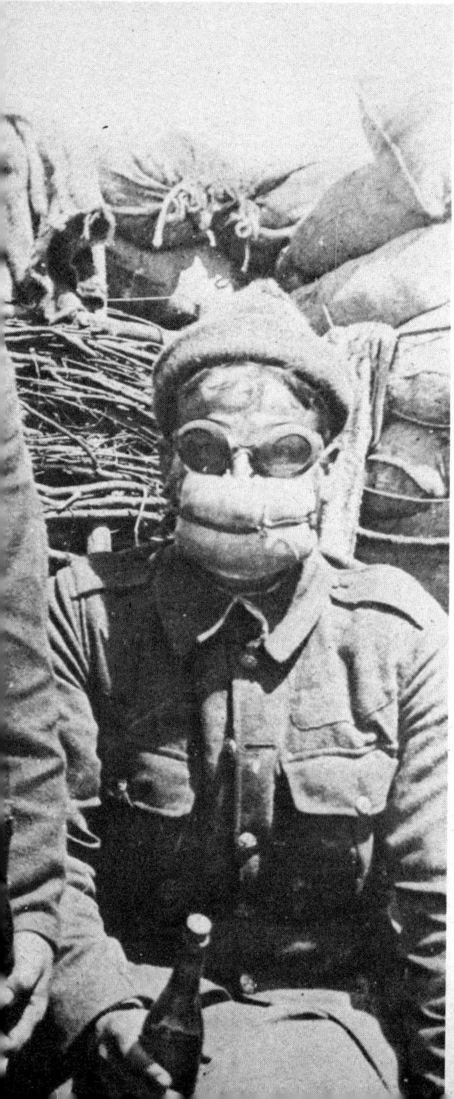

Meanwhile the Germans, realising that they had unleashed a form of warfare which, as had been foreseen by some, would rebound on them, began their own issue of face pads, soon to be followed by an absorbent cloth mask soaked in a sodium solution and which fitted over the nose and mouth rather like an elephant's trunk.

But what they failed to do was to correctly assess the effect of their Ypres attack. Due to the Allies' quick reaction in plugging the line during the darkness, and due to their strict secrecy about the number of casualties caused by the gas cloud, the real effect of the attack was not known by the Germans, and the High Command decided that while cloud attacks could continue they would not necessarily form part of a coordinated attack but would be simply a means of harassing the Allies. Another feature which tended to cool German enthusiasm for cloud attacks was the very basic fact that the prevailing wind stream in France and Flanders was from west to east, and only rarely did it blow the other, more convenient, way. This meant that the Allies could mount a cloud attack almost whenever they felt like it while the Germans were always waiting for a favourable wind.

For the moment though, the Germans were in no danger; there was very little the Allies could do in the way of retaliation, since they had no facilities for gas production comparable with those available in Germany; as we shall see, this was a fortuitous by-product of the way the pre-war heavy chemical industry had developed. The German organic chemical industry had grown by a vigorous policy of applying scientific investigation to commercial enterprise; money for research was always forthcoming, and on the commercial side no oppor-

Following the surprise gas attack at Ypres the French are issued with masks

tunity was lost to establish a world-wide market which, eventually, amounted to a monopoly. Dyestuffs and chemicals were sold at cut prices on overseas markets until local companies gave up the struggle and either closed down or sold out to the German combines. Thousands of patents were taken out to cover every possible product and process and freeze out the opposition; indeed, some patents were deliberately used to discourage research into various fields. As a result of all this activity -- which, it must be stressed, was carried on entirely for commercial ends and not, as has been suggested since, as part of some nefarious military plot -- in 1914 the vast majority of the world's organic chemicals and dyes came from Germany. Since almost all war gases involved production and research methods akin to those used in these industries, it followed that only Germany had the scientific plant and staff capable of producing war gases in any sizeable amount. France and Britain, and later the United States, had first to create their chemical plants or enlarge what few they had, before they could undertake any sort of gas offensive.

Whether it was the matter of the prevailing wind or the promise of better results is not clear, but part of Professor Haber's pioneers, reinforced by 25,000 more specially-made cylinders, now moved across Germany to the Eastern Front, and on 2nd May 1915 opened 12,000 cylinders to release 263 tons of chlorine against the Russians in the same Bolimov position that the T-Stoff shells had failed against. The result this time was far more significant; a death toll of 6,000 Russian troops and the breaking of the Bolimov position at last. Three cloud attacks were made in all, and the Russian casualties are believed to have exceeded 25,000 in the end. During May also, the other section of Haber's pioneers made more cloud attacks against British troops on the west, in the area of Loos; it reflects the ease with which chlorine could be defeated by even the primitive face veils when it is seen that the total British casualties from 330 tons of chlorine amounted to 7,000, of whom 350 died.

The success of the gas clouds, even if imperfectly appreciated by the Germans at first, led them to contemplate using gas in artillery and trench mortar projectiles once more, since these promised greater flexibility of use than did the cloud system with its reliance on wind direction. By June 1915 trench mortar bombs filled with methyl chlorosulphonate were issued; this was a lung injurant and lachrimator. But by a strange coincidence the French had hit on almost the same thing, ethyl sulphuryl chloride, and had begun fitting it into artillery shells. From now on new compounds came into the battle as fast as they could be developed and produced, either in the hope that they would be more effective than the last or in the hope that they would catch the other side by surprise and defeat whatever type of protection he was using. But of course, if you were about to use something new, it paid to be prepared, just in case the wind blew it back at you, so the first thing you did was to make sure that your own troops were adequately protected. In this way there grew up the practice of raiding the enemy lines to try and capture a gas munition or a respirator. Examination of either by experts would tell a lot; a skilled analyst given a respirator could often deduce what it was designed to protect against, and if your side didn't have that product in use it was a fair assumption that the enemy did and were about to use it on you.

As an example of the advantages of forewarning, the next German cloud gas surprise will serve. As early as July 1915 British intelligence agents

A German gas plant in occupied Poland

learned that the Germans were contemplating using phosgene gas later in the year. Phosgene is ten times more lethal than chlorine, and a surprise attack with it would have had worse effects than the Ypres cloud, since the current British protection, the 'Hypo Helmet' was only proof against chlorine and T-Stoff. The 'Hypo Helmet' was little more than a flannel bag, soaked in sodium thiosulphate (photographers 'hypo') and with a celluloid window in the front. As a result of this warning from the stealthy men, the 'P Helmet' was produced, similar to the 'Hypo Helmet' but with eyepieces and a rubber outlet valve which made breathing easier, and with the flannel soaked in a mixture of caustic soda, phenol and glycerine; the soda and phenol combined to form sodium phenolate which neutralised phosgene. Nine million of these helmets were made, and issues began in ample time to outfit all the troops on the Western Front by the end of November 1915. As a result, when the German phosgene cloud was released near Pilckem on 19th December, although eighty-eight tons of the gas were discharged, there were only 1069 casualties of which 120 died.

The 'P Helmet' was less effective than had been hoped, but help now came from an unexpected direction. Although no phosgene had been used on the Eastern Front, the British had warned the Russians of its likely appearance. Now the Russians came back with the information that they had been doing some experiments and had found that a substance called hexamethylene tetramine neutralised phosgene very effectively. Britain now added this substance to the brew in which the gas helmets were doused, calling the result the 'PH Helmet'. This could withstand high concentrations of phosgene for up to twenty-four hours, and it was probably the ultimate in the all-enveloping impregnated type of helmet.

The development of artillery shells was, however, a difficult business,

A French trench mortar firing a gas projectile

because there is rather more to it than simply taking a standard pattern of shell and pouring liquid gas into it. The gas had to be distributed at the target, which demanded an explosive charge, and, most important, the chemical substance had to be kept from contact with the steel or iron shell body, since all the early tear gases were highly corrosive.

It will be recalled that the T-Shells used a lead cylinder to contain the gas; fortunately the standard German method of shell construction lent itself to this system. The majority of German artillery shells were designed to be loaded with pre-formed blocks of high explosive, and thus either the head or the base could be unscrewed to give access to the body of the shell at its widest internal section. The 15cm T-Shells had a removable base, and after the 1.5 Kg filling of TNT had been set in place in the nose, the lead cylinder was inserted and cemented in place and the baseplate screwed in.

With the adoption of trench mortars as a gas-discharging weapon, the first German efforts were so crude as to hardly deserve the name 'projectile'. Their 9cm mortar bomb was simply a sheet iron tube with an iron base, into which two one-kilogram glass bottles of bromine were fitted, embedded in sand. The top was then closed by a wooden disc pierced with a hole through which ran a length of fuse to a burster charge of gunpowder above the top bottle. When this contrivance was fired from a mortar the safety fuse was ignited by the flash from the propelling charge and burned through to fire the burster at about the time the bomb arrived at its target.

This was soon replaced by a rather better product, but the pattern of

Varieties of British protection against gas, 1915

The P-Helmet is issued to all troops on the Western Front by November 1915

German projectiles remained generally the same -- a lead canister inside the shell, plus a powerful explosive burster charge which both distributed the gas and also shattered the shell into fragments.

The Allies, on the other hand, had a much more difficult problem due to their particular method of high explosive shell manufacture being totally different. Both Britain and France forged their shell bodies with solid bases and tapering heads, so that no solid object, such as a lead cylinder, wider than the fuse hole could be inserted into the shell. The only type of shell which differed from this construction and which could thus be readily adapted was the shrapnel shell, since this had a removable nose, and the French began their artillery shell programme by manufacturing two lead containers, semi-circular in section, which could fit side by side within the body of a 75mm shrapnel shell. One container was filled with tear gas and the other with a smoke-producing chemical -- usually stannic chloride. The use of this gave a smoke cloud for assisting observation and also helped to stabilise the gas and increase its density so that it had less tendency to rise from the ground to a harmless height.

This was a long and involved method of construction; it meant slow production and it was expensive in lead, while the use of the twin lead containers in such a small shell meant that it was heavier than normal, demanded a lighter propelling charge and had less range. Obviously something better was needed. The British approached the problem from a different angle; if the tear gas in common use at that time (ethyl bromoacetate was the French choice) was corrosive, then it might be better in the long run

A gas sentry with the new P-Helmet, introduced to protect against phosgene

Irish guards during respirator drill

to find a non-corrosive filling instead of going to great lengths to design a corrosion-proof shell construction.

The French however were determined to find a method of protecting their shell bodies so that the standard one-piece forged high explosive shell could be filled with gas, and they were of the opinion that solving this problem might very well be quicker in the long term than trying to find a non-corrosive filling. Throughout the latter part of 1915 they experimented in many directions, lining shell bodies with a variety of paints and varnishes, coating them internally with molten lead, coating them with lead by electrolytic action, trying various chemical treatments to produce an impervious finish, all without success. Then, at the end of the year, they perfected a method of blowing tough blue glass into the shell to produce a highly resistant lining, and finally

Various denominations of shell on display in a factory in England

German soldiers demonstrate the fitting of the fuse to a gas shell

solved the problem. The liquid gas could now be run straight into the shell through the fuse hole, and then a central enamelled tube carrying the smoke producer and an explosive burster was screwed into the mouth of the shell, a lead and asbestos washer between the tube and the shell body serving to seal the joint.

That solved the problem of lachrimatory shells for the French, but what they now needed was a good lethal gas shell. In June 1915 they tried shells filled with 'Sulvanite' -- ethyl sulphuryl chloride -- and in September, for the offensive in Champagne, with 'Clairsite' -- perchlormethyl mercaptan. Both these substances were lung injurants, but neither was particularly effective; Clairsite was only one-sixth as lethal as phosgene, it readily decomposed, and it had a pronounced smell which betrayed its presence long

before it could have any toxic effect. Since the results in the field did not justify the effort, the French stopped manufacture and turned back to the laboratory to produce something more effective.

After more trials, two fillings were developed which became standard with the French army for the rest of the war; 'Collongite' and 'Vincennite'. 'Collongite' was a mixture of phosgene and stannic chloride which took its name from the factory at Cologne which manufactured the phosgene. 'Vincennite', similarly named for the producing factory at Vincennes, was an involved mixture of hydrocyanic acid, arsenic trichloride, chloroform and stannic chloride.

The Collongite was quite successful, though it might well have been more effective if the French had not adhered to their policy of mixing the gas with a harmless smoke producer. It performed no useful purpose, since the phosgene was quite heavy enough to remain on the ground without assistance; it reduced the effectiveness of the shell by some 40 per cent by displacing phosgene; and its function as an observation aid could have been achieved by using other compounds in smaller quantities.

The Vincennite shell, however, was the greatest mistake the French ever made. Hydrocyanic acid is more popularly known as prussic acid and, there is no doubt about it, is one of the most virulent and rapid poisons in existence; at one time it was highly fashionable in detective novels, and the astute reader only had to see a reference to the 'smell of bitter almonds' to know what was coming next, though in fact the smell is somewhat indefinite and changes according to its dilution and the length of the period of exposure. I well remember an incident many years ago when a companion dropped a litre bottle of it onto the laboratory floor; we didn't stand around sniffing it, but en route to the nearest exit I noticed it didn't smell anything like almonds. What it does do is paralyse the nerves of the nose and palate, and after a few seconds one smells nothing at all, which is a useful feature in a war gas. But in order to be lethal the concentration has to be quite high; up to a certain level the human organism tolerates and eliminates the poison as fast as it is absorbed and no harm results. As one wartime writer put it, it was a 'kill or cure' compound; either you got enough to kill you or you were not affected at all.

The chemical characteristics of prussic acid are such that it is very volatile and it is thus extremely difficult to obtain and maintain a lethal concentration in the open air. This was the reason for the admixture with the arsenic and stannic chlorides; they formed a more persistent smoke which helped to retain the prussic acid gas in a concentration. But since they also reduced the chemical stability, chloroform had to be added to prevent the gas decomposing; this last additive was also of assistance in the filling process, since it kept the prussic acid gas dissolved and thus less dangerous to the operatives, but evaporated after the shell burst and released the gas to take effect. In any case, the final result apparently satisfied the chemists and Vincennite shells were produced in huge quantities, first being used in 1916 in support of the British attack on the Somme.

But Vincennite was a failure; the lethal concentration was rarely achieved and German casualties from it were negligible. It seems likely that one of the factors leading to its adoption was the fact that the German respirator of early 1916 afforded no protection against cyanides; but the Germans managed to learn of the French production programme, and before the shells reached the front had produced an additional filter for the canister of their masks which effectively countered the Vincennite when it arrived.

Inscribed gas alarm bell in the trenches

Duck

The British army made a number of trials with Vincennite and abandoned it as being worthless, while the Germans dismissed it without even bothering to perform trials. But the French persisted, filling over four million shells and using them throughout the war; it seems to have been their one blind spot as far as chemical warfare was concerned.

Although the Vincennite shell may have been a failure due to its filling, it has to be said that both this and the Collongite shell were design innovations of the highest importance, and went far to redress the balance over the Vincennite farce. The principal defect with artillery shell until these appeared was that the detonation of the high explosive had ill effects upon the gas; in some cases the heat and shock decomposed the inherently unstable chemicals so that the filling lost its potency. In all cases the detonation threw the gas far too widely, atomising some of it and driving a large quantity into the ground. This may well have been due in the first instance to the German insistence on maintaining a high fragmentation effect in order to adhere to the letter of the Hague Convention and produce a projectile with more splinter than gas effect; and having obtained a satisfactory design, they stayed with it, probably from sheer inertia -- it works, why change it?

The French though, when they came into the shell design business, had no particular need to consider the Hague Convention. Gas had arrived, for better of worse, and the time for legal niceties was long gone; the only consideration before them was that of efficiency. So the new French shells reduced the amount of explosive until it was just sufficient to split open the shell and allow the contents to escape, relying on the volatility of the gas to distribute it. This gave higher concentrations, since the gas was not so

Masked machine gunners in action on the Somme battlefield

widely dispersed by the detonation, and it also, as a bonus, made a good deal less noise when it burst, so that the arrival of gas shells could often go unnoticed in the general background noise of the battlefield.

In spite of the French army's lack of success with prussic acid, the cyanide group of chemicals still exerted a considerable attraction to the chemists and now the Austrians, who until this time had been content to obtain their gas requirements from Germany, were the next to look into this area. In September 1916 they attacked the Italians with shells filled with cyanogen bromide combined with bromoacetone and benzene. The effect, as far as lethality was concerned, was much the same as that to be obtained with hydrocyanic acid, though with this mixture even higher concentrations were required. But in addition it was strongly irritant and lachrymatory and a good deal less volatile than the French Vincennite mixture. For all these virtues, it was not particularly successful; it tended to decompose during storage and gradually change into a fairly harmless substance, so that unless the shells were fired soon after filling, they had less and less effect as time went by.

A similar compound, cyanogen chloride, was tried by the French under the name 'Vitrite'. It was mixed

with arsenic trichloride to make it heavier and to delay evaporation, and its assets and defects were much the same as those of the bromide version. It seems not to have had much success and eventually both the French and Austrians abandoned the use of these compounds.

British development of artillery shell was slower than that of France or Germany principally because they had elected to give first priority to developing a cloud gas capability, and second priority to developing a suitable gas projectile which could be fired from a trench mortar. The invention of the Stokes 3-inch mortar in 1915 had provided an excellent weapon for the rapid delivery of explosive bombs and the idea of using it as a gas delivery system was very quickly examined. But the capacity of the 3-inch bomb was considered to be insufficient, and a request to Sir William Stokes soon produced a 4-inch model in September 1915. This had almost the same degree of mobility as the smaller weapon but the bomb could carry three times as much payload, anything between six and nine pounds of gas being carried according to the nature of gas used.

Britain's gas manufacturing capacity was practically nil when gas war-

Austrian infantry on the Italian front

fare began; so that instead of taking whatever was convenient, they first investigated numerous substances and then decided on what they wanted to manufacture, setting up the necessary plant. Much of the laboratory work and development engineering was done by universities and private companies on Government contracts, and manufacture was largely contracted out to commercial chemical companies if their available plant was suitable, or by Government-built factories run either by private concerns as the Government's agents or as offshoots of the Ministry of Munitions.

The first substance to be produced in quantity was ethyl iodoacetate, known by the code letters 'SK'. The use of codes was adopted by all the combatants in an attempt to conceal the nature of the gas from the enemy until it was actually in use against him; obviously, such codes had to be obscure so as not to be capable of decipherment, although this rule rather went overboard in later years. The British system of allotting codes was, however, more obscure than most and very definitely in keeping with the British sense of humour. SK got its code from being developed at the Imperial Institute in *S*outh *K*ensington; others were even more far-fetched.

SK was one of the most powerful lachrimators used during the war, and, in addition, it was moderately toxic. A concentration as low as 1.5 parts of SK in one million of air was sufficient to be intolerable, and 160 parts per million was fatal after ten minutes exposure. A considerable advantage was the fact that it did not attack iron or steel and could therefore be loaded into shells and bombs without going through all the troubles the French had suffered in trying to develop a corrosion-proof shell lining. The first use of SK was in shells for the 4.5 inch howitzer at the Battle of Loos on 24th September 1915, but this was a cast iron shell with poor capacity, and thus a sufficiently heavy concen-

tration of the gas could not be obtained. As a result, the manufacture stopped until a better design of steel shell with more capacity could be produced. The supply of SK was then diverted to filling 4-inch Stokes mortar bombs, and these were much more effective, carrying a better percentage of gas and being capable of delivery at high rates of fire, thus producing effective concentrations.

It can be seen that throughout British development and employment of gas there ran a thread of reasoning which was fundamental to their system. They had, by this time, been recipients of a variety of gas attacks by cloud, artillery fire and trench mortar fire, and analysing the results they had come to a definite conclusion; that the best effect could only be obtained from gas by a system of delivery which produced the highest possible concentration in a specific area in the shortest possible time. In short, success could be guaranteed if the enemy were swamped with gas at such an over-

The explosion of a 4-inch Stokes mortar gas shell on the proving range

whelming rate that it took effect before he had time to take precautions.

Another fundamental of British thinking, in which they were in advance of the Germans, was in the whole basic question of the purpose of gas. To the Germans, gas was an adjunct to manoeuvre, to be used in support of an attack, and it was rarely used otherwise. To the British, on the other hand, gas was a weapon to be used whenever there was a likelihood of catching the enemy unawares, a weapon of harassment and attrition. Of the various methods available to the British the cloud seemed to be the one most effective and most in line with their thinking, but it was too dependent upon wind and it gave too much warning of its approach; moreover it devoured a lot of time and manpower in preparing it. Therefore what was wanted was a system of delivery so rapid that it amounted to depositing a cloud in the required place without having to rely on the wind to drift it there.

A Lieutenant W H Livens was involved in the organising and operation of the British cloud gas attack at Loos, and he reasoned that the better solution would be to get the gas cylinder among the enemy before opening it to release the gas. Before he could do anything with his ideas he was put in charge of a flame-thrower company during the Battle of the Somme in 1916, and while so engaged he developed a primitive device for throwing water-cans full of flame thrower fluid onto the enemy positions by using an oil-drum as a form of mortar. From this humble beginning he began to consider the possibility of fitting some sort of burster to the standard gas cylinder and launching it from a mortar. But the standard gas cylinder weighed about a hundred pounds, and launching such a weight from a simple device wasn't so simple after all. Eventually his ideas crystallised into the 'Livens Projector', a simple muzzle-loading mortar of eight inch calibre. Batteries of these were dug into the ground behind the trenches and loaded with an electrically fired guncotton cartridge and a bomb which was nothing more than a short gas cylinder with a burster charge and simple fuse. These bombs held about 30 lbs of gas; once installed and loaded the propelling charges were wired up in series and the whole battery was fired at once.

The first Livens Projectors were fired at Beaumont Hamel in October 1916, more or less as a trial; their effect was satisfactory, even though the weapon was in primitive form and needed perfecting, and arrangements were put in hand for mass production of projectors and bombs. This initial outing appears not to have been appreciated by the Germans, and it also seems to have escaped the notice of the historians; the 'official' date for the first use of Livens Projectors is usually given as 4th April 1917, but Brigadier-General Hartley, who was Assistant Director of Gas Services at GHQ, referred to the Beaumont Hamel operation in a lecture he gave in 1919.

Having proved on a small scale that the weapon was workable, the decision was taken to hold it back until sufficient projectors and bombs were available to bring it into the field in worthwhile numbers, and then not to use it until it could be decisively employed during a major operation. Once unveiled it could then be used as a weapon of attrition, but it would have been foolish to throw away its surprise value by employing it on day-to-day work. The lesson learned on the Somme, where the tank was thrown in on the wrong ground and in too small numbers, was fresh in many minds, and the Livens Projector promised to be too good a weapon to be squandered in similar fashion.

Lieutenant W H Livens' invention, the Livens Projector

EXPLODER
LIVENS
Bomb

Loading charges into Livens Projectors, first fired in October 1916 at Beaumont Hame

Retaliation

German prisoners captured by the Canadians during the Battle of Vimy Ridge

On Easter Monday 9th April 1917, the British and Canadian armies attacked at Arras, to take Vimy Ridge and punch a hole in the Hindenburg Line in an assault which was, in its initial stages, one of the most successful battles of the war. As was, however, usual in those days, the reserves and cavalry who were to have poured through the broken German defences failed to materialise and the exhausted infantry could do no more than halt and form a new line.

General Allenby, the commander of the operation, while a taciturn man who found it difficult to communicate his ideas to others, was nevertheless a very able general and one who was particularly alert to the chance of surprising his enemy with something new. He was also alert to the capabilities of gas when well used, and for the attack at Arras he decided to use not only artillery but the new Livens Projector as well. Vast amounts of artillery shells filled with chloropicrin, a powerful lung injurant known in British service as 'PS' (because it had been developed in the Lever Brothers' laboratory at *P*ort *S*unlight, Cheshire), were now available, together with quantities of Livens bombs filled with 'CG', the British code for phosgene (derived from the French factory at Collonge, where the first Allied supplies were made).

Getting the Livens Projector into action was a simple but laborious task: it was not like a mortar, poised on a baseplate and bipod. The design was far too basic for that, since all the projector consisted of was its wide barrel with a reinforcing plate welded to the bottom; no bipod, no sights, no adjustments of any kind, just a tube. It was emplaced by digging it into the ground; a long trench of V-section, four feet deep and with the sides angled at 45 degrees, was excavated; the projector barrels were then laid with their bases in the bottom of the trench and their muzzles pointing towards the enemy, just above the surface of the ground. The earth was then shovelled back on top to hold them securely in alignment. One trench would hold anything up to forty projectors, and, naturally, preparation was an activity which was carried out behind the front line trenches and concealed from the enemy's view.

When the time to fire drew near, the charges of guncotton were placed in the barrels, and their electric firing leads run out from the muzzle. Then the bomb was inserted, the firing wires linked up and tested for continuity, and connected to a firing dynamo. At the last moment the firing officer

Muzzle caps on a Livens Projector emplacement

locked the dynamo, took the key in his pocket for his own peace of mind, and then went down the line of projectors pulling out the safety pins from the bomb fuses. Then he returned to his dynamo, unlocked it once more, lifted the plunger, and at the appointed time rammed it down to fire all the projectors at once. A roar, a blinding flash, and the bombs were hurtling through the air, turning end over end, trailing sparks from their burning fuses, to land with a thud forty-five seconds later and a mile away.

At Arras no less than 3827 projectors threw fifty-one tons of phosgene into the German trenches in one colossal salvo. A German prisoner afterwards reported that the enormous flash from the projectors was assumed to be an Allied ammunition dump exploding, and great was the merriment thereat. The incessant drumroll of gunfire common to the Western

Front masked the sound of the bomb's flight and arrival, and the first indication of anything untoward was the sudden envelopment of the German lines in a cloud of phosgene. The concentration was so great that one deep breath was sufficient; over a hundred men died instantly, before they could put on their respirators, and about 500 were hospitalised -- the exact figures have never been determined since the attack which followed immediately afterwards rather confused the returns.

Meanwhile the artillery had been engaged in a 48-hour planned bombardment -- lengthened to 72 hours due to the attack being postponed -- and in the latter part of this they dumped 60,000 shells containing 150 tons of chloropicrin into German artillery positions and rear areas, a concentration which effectively hampered the

Livens Projectors deliver a salvo; the bombs can be seen streaking skywards

British cavalry at Arras await orders to move forward

German artillerymen in their attempts to beat back the attack.

An interesting fact emerges; the 150 tons of chloropicrin killed an estimated fifty men and injured another 600 or so -- an average of 516 lbs of gas per casualty; the Livens attack shows an expenditure of 190 lbs per casualty. Yet when the British had used phosgene in a cloud attack on the Somme, it had produced 6500 casualties for an expenditure of 60 lbs of gas per head, and when the Germans first used chloropicrin their average was 395 lbs per casualty. The answer, of course, is simply that by this time gas was less of a novelty: protection was steadily improving, and while a new gas might achieve some degree of surprise when it was first used, the likelihood was that provided the troops were sufficiently alert to don respirators, then the casualty list would not necessarily be a high one. But it must be borne in mind that even if the use of gas failed to achieve even one casualty, it was still managing a useful amount of disruption; for while the measures for individual protection were in most cases adequate, they were very far from convenient.

As we have already seen the first protection was no more than a wad of surgical cotton or lint soaked in absorbent or neutralising chemicals and tied across the mouth and nose. This was avowedly an emergency measure, but it was effective to some degree against chlorine. The arrival of lachrimatory gases made it necessary to protect the eyes as well as the lungs, and to this end goggles were issued to be worn in conjunction with the face pads. Then began the period of the all-enveloping 'Gas Helmets' already described, sacks of treated flannel which the soldier pulled over his head and tucked into the neck of his tunic, breathing through the cloth itself. The principal drawback was that the exhaled breath caused a build-up of carbon dioxide inside the helmet, and this was rectified by an improved model which had an exhaust valve.

But how do you suppose it felt to have to march, fight, shoot, run and even sleep with your head in a flannel bag soaked in hypo glycerin and washing soda? The celluloid windows or eyepieces soon clouded over, grew dim with age, cracked, and generally made vision difficult; if the weather was warm, you sweated; if it was cold the exhaled breath condensed on the inside of the bag and ran down your neck -- taking with it a proportion of hypo and washing soda. Even if you were immune to the gas around you, it could hardly be said that you would be at peak fighting efficiency. And this, of course, was how gas could have an effect without necessarily causing casualties; by forcing the enemy into his mask, it reduced his fighting efficiency. On more than one occasion the British army fired smoke shells at the Germans, lacing the mixture with a handful of tear gas shells; quite a harmless mixture, but how were the other side to know? The French on one or two occasions fired shells containing strange but quite innocuous liquids into the German lines; until the contents had been collected, sent back to a laboratory and analysed, nobody could be sure that it was not some new form of deadly poison. There was only one course open in either case; to wear the respirator, and once into respirators, the defences were proportionally hindered.

As a result of this, there was a constant search for a respirator which would be effective but which would be less fatiguing to wear for long periods. The German army moved straight from the face pad to the self-contained canister respirator. During the latter months of 1915 a mask consisting of a leather facepiece with eyepieces and a metal canister was issued. The facepiece was impregnated with tallow and oil to make it gas-proof, the eyepieces were compound, with an outer glass lens and an inner lens of celluloid

A frying pan serves as an alarm signal in a German trench

Above: Royal Field Artillery gunners wear masks as protection against German gas shells. *Below:* German heavy howitzer detachment in action. *Right:* The Germa mask issued in 1915

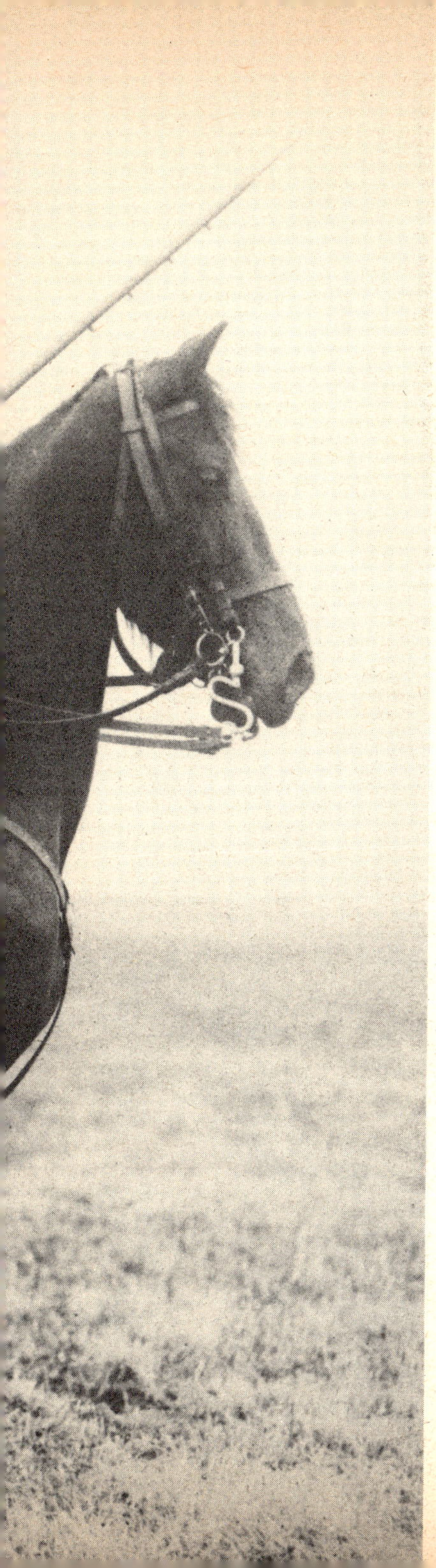

treated to prevent misting, and the canister contained layers of absorbent materials -- charcoal, pumice and kieselguhr -- mixed with various chemicals to filter and absorb the gases. While this was more comfortable and effective than the face pads, it had a slight drawback in that the facepiece and canister had to be carried separately and screwed together when required, though this was later remedied in an improved model which was one-piece and could be put on instantly.

The French produced a mask made of thirty-two layers of muslin, chemically impregnated, through which the wearer breathed, and this fitted on the face rather like a horse's feed-bag. Celluloid eyepieces were provided, but these invariably misted up within minutes of putting the mask on. They then went to the other extreme and produced the 'Tissot' mask, a luxurious piece of work with a moulded pure rubber facepiece and a flexible pipe running to a large container of absorbents and filtering agents carried on the wearer's back. It had an exhaust valve sticking up between the wearer's eyes, and was cleverly designed so that the inhaled fresh air passed across the eyepieces and kept them free from mist and condensation. It was a splendid device, quite the best of the war, but it was far too cumbersome for the line soldier and was generally used by artillerymen and other troops who were not expected to carry the infantryman's 60 lbs of miscellaneous gear on their backs while at work.

The British now turned to a canister solution and produced the 'Tarbox' or, more properly, 'Large Box' respirator. This was originally designed to be used by troops of the Special Brigades responsible for operating gas munitions and who might be expected to have to meet high concentrations of gas from time to time. A facepiece of

German dispatch riders are given masks in case of having to pass through pockets of gas

A German anti-aircraft crew poses in gas-attack garb

A gas protection box for German carrier pigeons

Early French gas mask

An improved French gas mask

The French 'Tissot' mask

twenty-four thicknesses of muslin soaked in absorbent chemicals was connected to a large canister of charcoal, soda lime and potassium permanganate carried on the chest. But since this facepiece was only marginally effective, the wearer had to put a clip on his nose to prevent him breathing through it, and hold a mouthpiece, forming the end of the rubber hose connecting facepiece and canister, in his mouth, breathing in and out through the canister. Moreover the facepiece only covered the nose and mouth; goggles still had to be worn to counter lachrimatory gases.

The designs of the British and German masks of this period exhibit an interesting difference in priorities; the German mask was primarily intended to be efficient, while the British mask was not merely to be efficient but had to have a canister large enough to absorb gas for long periods and, what was more important, allow the wearer to inhale the volume of air needed when involved in strenuous physical activity. While the German mask was undoubtedly efficient, the size of the filter canister was restricted by considerations of its weight; if it were too large and heavy it would pull the mask off the wearer's face, or at least make him tired of carrying it. Making it small enough to be worn comfortably brought the size down to the point where a man involved in heavy work -- digging, loading a heavy gun, or even trying to run across No Man's Land -- was soon brought to a stop since he could not inhale the air his lungs demanded.

Wartime and immediate postwar statements by German authorities tended to gloss over this aspect, and gave the opinion that the German mask was the more compact and efficient, and that the Allies adopted separate canister patterns because of their deficiency in chemical knowledge which prevented packing the desired protection into a more compact space. But the truth of the matter lay, in fact, in Germany's acute shortage of rubber; in spite of considerable efforts to produce a synthetic substitute, this shortage bedevilled them throughout the war. It was for this reason that the facepiece of the German mask was of leather, and it was for this reason that their mask had to be a one-piece unit; there was not sufficient rubber to manufacture the millions of flexible tubes which a two-unit respirator would have demanded.

However, the British 'Large Box', designed as it was for specialist use, was really too efficient for normal use, the canister being over large for the task it was normally called upon to perform, and a fresh model the 'Small Box' was soon produced for general issue in April 1916. This brought the mask to its final form, the form which, with minor modifications, remained the standard for many years. The rubber-cloth facepiece covered the whole face and contained eyepieces; while it was quite tightly fitting, the mouthpiece and nosepiece were retained for additional security, just in case the Germans came up with a gas that could penetrate the facepiece at some future date. The canister contained layers of charcoal, soda lime, and potassium permanganate, and the general result was a highly efficient article of protection. Coupled with a constantly improving standard of 'Gas-Discipline', the military phrase covering such things as gas sentries, alarm systems, and training in detection of gas and use of the mask, the British (and later the American troops who also adopted the Small Box Respirator) were amply protected against all the gases so far discovered.

Once given that one's opponent has developed a complete system of protection against your latest weapon, there are two courses open: you either give up using it entirely, or you go back to basic research and redevelop the weapon so as to overcome or circumvent the protection. And this is

The standard British mask, 'Small Box'

what the Germans now did. The British and French masks could cope with chlorine, phosgene and the various other compounds in use; therefore it became imperative to discover a compound which could defeat the masks, otherwise the advantages of gas warfare were cancelled out. Either a gas had to be found which would penetrate the mask, or a gas had to be found which would ignore the mask and attack some other part of the body.

In fact the German chemists solved both puzzles, and developed two totally new approaches. The first of these, and the one to achieve the greatest notoriety and success was revealed in an artillery bombardment on the night of 12th July 1917 against British troops near Ypres. The only sign of anything untoward was a faint smell, likened by most observers to horse-radish. No tears, no interference with breathing was observed, and since the faint smell bore no resemblance to any known form of gas, no alarm was sounded. It was some hours before the symptoms began to appear: inflammation of the eyes, vomiting, and, most serious, blistering of the skin. Mustard gas had arrived on the battlefield.

Dichloroethyl sulphide, which became known as mustard gas to the British and Americans, Yperite to the French and 'Lost' to the Germans (from the initials of the chemists who developed it) was the first vesicant substance to be used, and by its property of attacking the skin it bypassed the mask and inflicted casualties in spite of the existing degree of protection. In addition, its relatively weak smell, even in lethal concentrations, gave a false sense of security and led to large numbers of casualties among men who ignored the faint aroma in the assumption that if it was a war gas then there wasn't enough of it to be dangerous. These two proper-

A Portuguese gas sentry at Neuve Chapelle

ties made mustard the most effective war gas developed for many years. Its effect on troops in the first days of its use is well described by the account of one of its victims:

"Going up (to the line) we met a terrible strafe of high explosive and gas shells in Nieuport..... the gas was clearly visible and had exactly the same smell as horse-radish. It had no immediate effect on the eyes or throat. I suspected delayed action and my party all put their masks on.

"On arriving we met a thick cloud of the same stuff drifting from the front line. As it seemed to have no effect on the eyes I gave orders for all to put on their mouthpieces and nose-clips so as to breathe none of the stuff and we carried on. Next morning myself and all the eighty men were absolutely blind. One or two never recovered their sight and died."

Standards of 'Gas-Discipline' improve; troops enter a gas trench at a training school

French troops get used to wearing a more sophisticated mask

In all, between 12th July and 4th August, the German artillery fired about a million mustard gas shells at the British army, causing an estimated 14,726 casualties of whom about 500 died. The most difficult problem for the Allies lay in the vesicant action, since the drops of gas would penetrate clothing, boots and even rubber to burn any area of the body. Furthermore it was a highly persistent gas, as already noted, and areas drenched in it were hazardous for long periods afterwards. This factor soon came to be recognised as a method of deducing where an attack would fall. For example, in April 1918, when the German army was preparing an attack to the north of Lens, there was no gas shelling between the La Basee Canal and Armentieres. There was, however, heavy mustard shelling south of the canal, and the shelling of Armentieres was so severe that at one stage eye

Gas drill for artillery horses

British troops under gas cloud attack

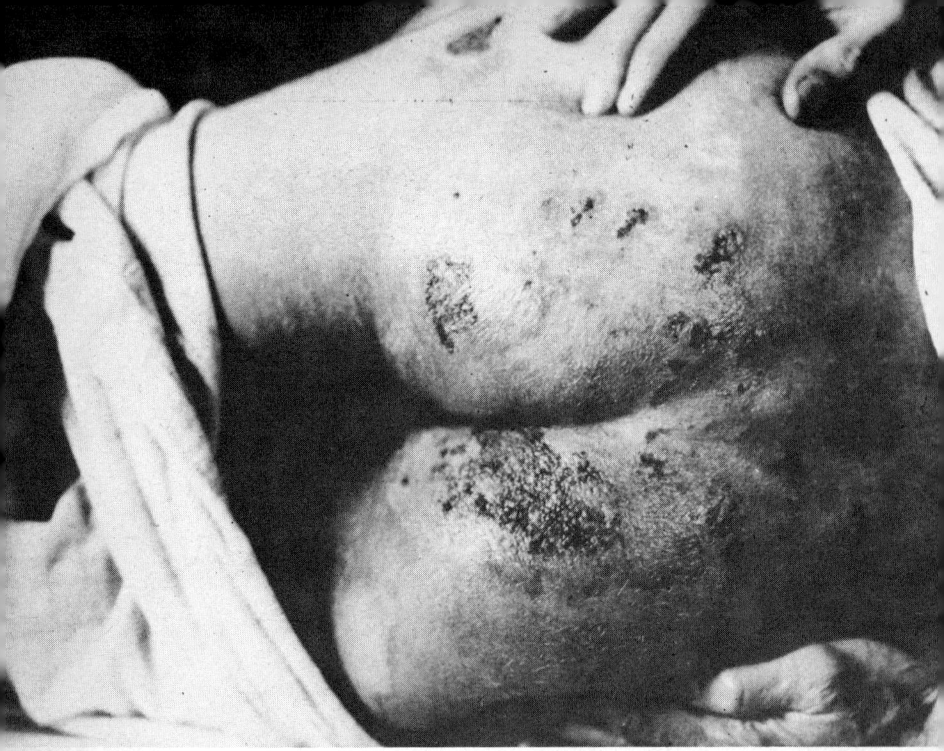

Blisters caused by mustard gas

witnesses reported that the gutters were running with neat mustard gas. It was deduced from this that the forthcoming attack would fall in the Portuguese-held sector of the front, which remained untouched by mustard, and, sure enough, on 9th April the entire weight of the German attack fell in the indicated area.

The second surprise prepared by the Germans was a means of attack which could actually defeat the respirator. As we have seen, the Allied respirators could absorb all the known gases; moreover, all the lung injurants so far used were relatively slow in action, so that if warning were given and the soldier was able to get his mask adjusted quickly, he was unlikely to be affected. Even if he did have one or two breaths of the gases, it usually took some hours for the effect to be noticeable, though in some cases it was lethal. The German problem was to develop a substance which would penetrate the Allied canisters and take effect rapidly.

The gases developed to solve this problem were the arsenic-derived sternutators, and their notable feature is that they are not, in fact, gases at all, but suspensions of finely-divided solid material. Since the canisters of the Allied masks were designed to absorb or neutralise chemicals, this fine smoke or dust cloud could penetrate through the interstices in the charcoal and soda-lime, and within a few seconds of inhalation gave the victim terrible headaches, nausea and caused violent and uncontrollable vomiting which made it quite impossible to continue wearing the mask. To take even greater advantage of this it became standard German practice to fire this type of gas -- known as the 'Blue Cross' group -- in combination with a lethal lung injurant. The Blue Cross penetrated the mask and forced its wearer to remove it, and his distress was such that he would not notice that he was also breathing quantities of the

Green Cross lung injurant gas.

The Blue Cross gases are often referred to as 'Arsenical Smokes' which is a more accurate definition than 'gas' for them. The first to appear was diphenyl chloroarsine, known by the Germans as 'Clark I', and it was introduced at the same time as mustard in order to provide a non-persistent agent for use in areas of attack. In addition to its immediate nauseating effect, since it was an arsenic compound it could, in sufficiently high concentrations, act as a systemic poison, but in practice such concentrations were difficult to achieve in the field and the usual effect of 'Clark' was to incapacitate the victim within two minutes. After resting in an uncontaminated atmosphere the victim gradually recovers and in fact very few deaths were attributed to this gas.

Diphenyl chloroarsine had one considerable drawback as a chemical agent, particularly in that place at that time; it was readily decomposed by water. In order to overcome this, an analogous substance, diphenyl cyanoarsine or 'Clark II' replaced 'Clark I' in May 1918. For all practical purposes the effect was the same, though in theory it was some 50 per cent more lethal.

The Clark smokes were highly effective; they took effect at such low concentrations (one part in ten million was enough for nausea to set in) that a handful of shells flung into an area would cause an alarm and make the troops mask, lowering their efficiency; as a harassing agent the arsenical smokes have never been excelled.

However, the Allies were spared the worst; in the first place, the British had already considered the possible use of 'particulate smokes' being used, and had issued an additional section to be fitted to the bottom of the respirator canister. This carried a 'mechanical' filter (as opposed to a 'chemical' one) of porous paper and cheesecloth which trapped most of the irritant particles. In the second place,

A mustard gas victim

the Germans were too optimistic and put into service an inefficient design of shell. The ideal method of dispersing these smokes is by heating them, when they volatilise into microscopic particles which are extremely difficult to filter. But this, of course, is impractical in a shell, and the German design used a cylindrical filling of high explosive with a central 'core' of Clark mixture. Thus when the high explosive was detonated the Clark would be pulverised and dispersed in fine dust. That was the intention, but it failed to work that way; the placing of the explosive around the Clark led to it being largely destroyed, the heat of the detonation was too brief to have the expected vapourising effect, and the dispersed material (what there was of it) was in too large a form for it to penetrate the canister filters. As a result, though the Blue Cross attacks were potentially a considerable hazard, in fact they proved to be much

British troops prepare, amid bursting gas shells, to defend their position

A member of the American Medical Corps examines a gassed soldier

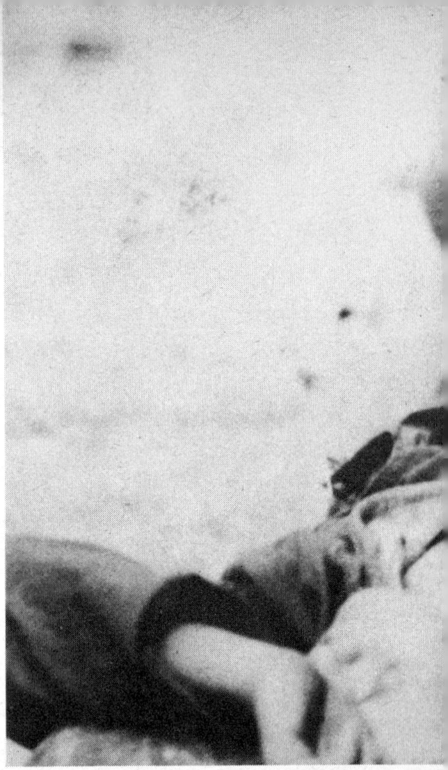

less dangerous than they might have been, and much less effective than the Germans believed. For example, the US army in France suffered only 577 casualties and three men dead from this class of agent. On the other hand the Russian army, with less effective masks and poorer gas discipline, suffered considerably. They were first treated to 'Buntkreufeuer' -- mixed Blue and Green Cross shell fire -- in September 1917 at Uxhüll on the Dvina River. Fire was directed at the Russian artillery positions covering the river crossing area, and after a two hour bombardment the guns were silenced and an estimated thousand were casualties, of whom about ten per cent died, victims of the Green Cross element of the bombardment.

American casualties at a field hospital after a gas attack

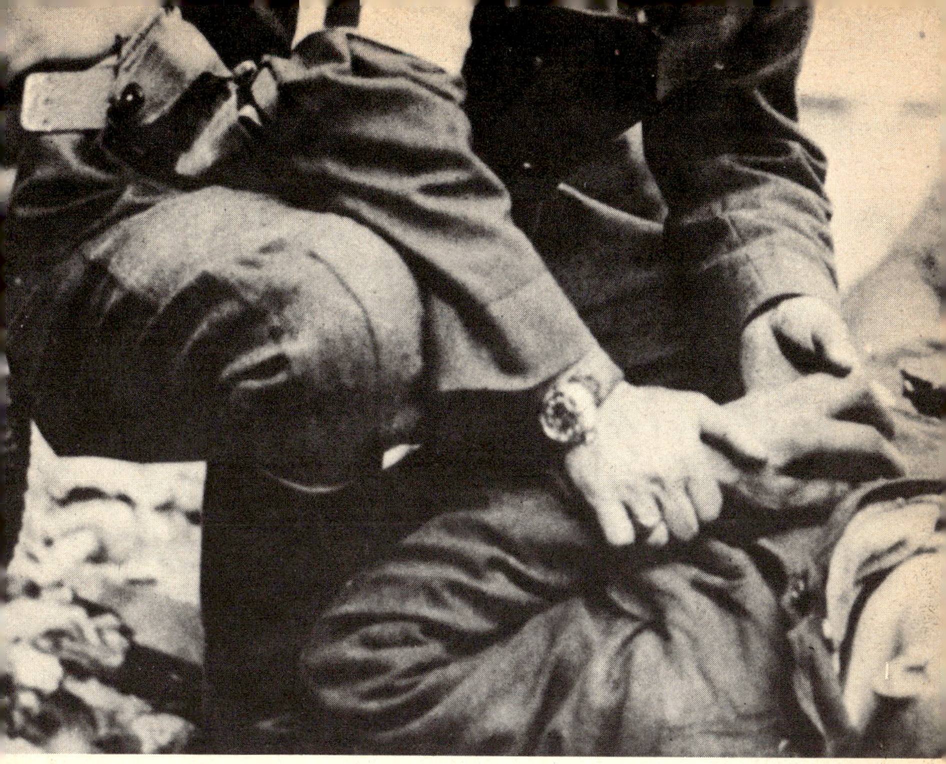

Organisation

A churchbell is pressed into service as a German gas alarm

So far in the story of the use of gas in the First World War, one fact seems to stand out: the fact that the Germans called the tune to which the Allies danced, and it is instructive to study the relative organisations on each side so as to understand why this was so.

The reason for the initial German superiority in this field had its roots in commerce. In the early years of the century, Germany had carved out a considerable niche in the organic chemistry industry, owing largely to the increasing knowledge of and demand for synthetic dyestuffs. Half a dozen companies had specialised in this field, and by price-cutting, amalgamation, buying out competitors and other well-known commercial stratagems, by 1914 they had what amounted to a monopoly of the production of dyestuffs and organic industrial chemicals for the world. Eventually all the companies merged into one organisation, the Interessen Gemeinschafte fur Farbenindustrie, commonly known as the IG Farben.

Thus in 1914 Germany had the highest concentration of chemists and chemical plant in the world, and their facilities for research and experiment were unrivalled, since the companies had always followed a policy of aggressive development, ploughing back profits into their plant and providing their extensive research staffs with whatever equipment they wanted.

The basic materials of the organic dyestuff industry at that time were coal tar, phenol, naphthalene and acids, and the significance of a healthy chemical industry to a country at war is realised when one appreciates that these were also the raw materials of the most of the explosives of the day. Moreover, many of the dye manufacturing processes were but a step removed from explosive processes; for example, a first step on the road to many colours used by the cotton-dying industry was the production of nitrotoluene. By changing the next step in the process, a relatively simple matter, TNT could be produced instead of cotton dyes.

As we have already seen, the German army and government were convinced of their ability to bring the war to a rapid conclusion; so convinced that they took no steps to mobilise the chemical industry for war production. But as soon as the needs became apparent, Germany was in a much better position than most other countries since the instrument was ready to hand. Owing to the centralised organisation of the industry it became a relatively easy matter to bring a specific substance into production; a request was simply made to the central office of the cartel (which, it should be added, did not actually become IG Farben until 1916) and from then on the directors of the individual plants took over, allotting production to the plant best suited to the process, arranging the supply of raw material, arranging the continuation of processing in another plant if that were technically more convenient, and finally shipping the finished article to wherever it was needed.

This was the system adopted for the manufacture and supply of all German war gases, none being produced in any Government-controlled plant. As an example, when mustard gas was demanded for the first time, the first stage -- preparation of ethylene chlorohydrin -- was carried out at the Badische Soda und Anilin Fabrik at Ludwigshafen, since plant for the production of this material formed part of their synthetic indigo manufacturing process. This material was then shipped to the Bayer factory at Leverkusen for the remaining stages to be carried out since the necessary skilled labour and specialised plant were more easily provided there and also because Leverkusen was considered less likely to be a target for Allied air raids.

The German chemical warfare organisation thus grew up in the shadow of the IG Farben. It began in December 1914 when Tappen was put in charge of the development and production of T-Shell and mortar bombs and Professor Haber was given the task of organising cloud gas techniques. After the first cloud gas attack in 1915, Haber was given the additional responsibility for devising a respirator for use by the troops, and work on this was begun at the Kaiser Wilhelm Institute, numbers of chemists being demobilised from the army and sent to work there. The protection of troops became part of the Army Medical Service's responsibility, and once the institute had developed a respirator the arrangement for its production was put in hand by the Chief Medical Depot in Berlin. Thus, by the latter part of 1915, the German gas warfare programme was a fragmented affair with what amounted to three different departments operating independently of each other.

Early in 1916 the success of the new French gas shells at Verdun led to a revision of German gas shell policies and designs, and the responsibility for design of new shells was passed to Professor Haber. By this move, he became the chief technical advisor for both the defensive and offensive aspects of chemical warfare, and the whole of the experimental work was carried out by the Kaiser Wilhelm Institute which was considerably expanded to cope with the extra load. Finally, late in 1916, the organisation was unified by removing the responsibility for defensive measures from the Medical Service and setting up a Chemical Section within the General War Department of the War Ministry, with Haber as its chief, to deal with every aspect of chemical warfare. Haber remained Director of the Kaiser Wilhelm Institute and this establishment was now organised as a military concern, every aspect of chemical warfare research being carried out there with the exception of actual field trials.

With the research and manufacturing sides organised, the only remaining question was the actual charging of gases into munitions. For cloud gas, this was left to the gas manufacturers, since they had the necessary plant for charging cylinders and could do it far more efficiently and quickly than any extemporised factory; moreover their rail connections allowed filled cylinders to be sent to the front whenever needed. It was artillery shell which caused problems; the T-Shell had been produced in stages. The lead cylinders were filled in a chemical works in Leverkusen and shipped to an ammunition assembly plant. The shell bodies were given their explosive filling in a normal explosives filling factory and also shipped to the assembly plant where the gas containers were inserted and the shells completed. Although cumbersome it was a satisfactory system where relatively small quantities were concerned and for this particular design. But with the introduction of shells filled directly with liquid a question arose of how long they would remain serviceable before deteriorating or beginning to leak, and the early Green Cross (phosgene) shells were filled in field filling plants set up behind the front as and when they were needed.

Two of these plants eventually expanded to become the principal shell filling plants, one at Mancieulles, near Audun le Roman, for the Western Front, and one at Warsaw for the Eastern, but their methods were comparatively crude and gave unsatisfactory results. As soon as it became apparent that a properly filled and sealed shell could be stored for long periods without leaking and without the contents suffering deterioration, it was decided to centralise production in a government-controlled factory in Germany.

The original intention was to find a suitable site near Berlin where all the sections of the Chemical Warfare Organisation (with the exception of the Kaiser Wilhelm Institute) could be installed, thus bringing into one complex all the technical experience and ability, and providing for a useful interchange of information. But there was no locality near Berlin considered safe, since it was intended to conduct large-scale field trials in the new establishment, and eventually an area on the Luneburg Heath was selected.

On this site, five kilometres from the large military camp of Munsterlager, the Brelch Gas Depot was built, a hundred-square-kilometre area which was to contain a complete filling factory capable of dealing with all requirements for shells, mortar bombs and cylinders, together with an artillery range, a chemical laboratory, a field trials unit, the Army Gas School and all the necessary barracks and services. With the introduction of mustard gas the plans were changed to provide for two filling factories, one to deal with Green and Yellow Cross shells which could be filled without the need for refrigeration plant, and the other to charge phosgene and other agents which needed refrigeration.

The intended programme never materialised; Professor Haber later complained that while he never lacked for anything he wanted for chemical warfare, when it came to funds for building Brelch he could never get enough money to allow construction to proceed quickly. As with most armies, while he was dealing with munitions and the men who needed them, he had little trouble, but once he came up against the construction and works departments he was up against a bureaucracy which had changed little since the pyramids were built. Due to this slow progress, and also to the fact that the gas manufacturing factories refused to undertake the charging of shells with mustard gas since they considered it too dangerous, an emergency filling plant had to be erected at Aldershof near Berlin, known as Field Ammunition Depot No 3, which was exclusively used for charging Yellow Cross shells. This plant could turn out 24,000 77mm field gun shells each day and was staffed by 3,000 soldiers who were graded as unfit for active service at the front.

Except for the slow construction of Brelch -- it was never completed to its intended plan -- it can thus be seen that the German gas warfare organisation was a relatively well-integrated and smooth-running affair. The Kaiser Wilhelm Institute would determine a suitable chemical agent: the IG Farben would organise its production within their organisation in the most expeditious and efficient manner: and the filling and charging factories would produce the actual munitions and send them to the front. It might be added that this functioning organisation was overlaid by a bureaucratic set-up which purported to control it, but as one post-war comment had it, "All they provided was plenty of paper and cripples."

Now let us cross No Man's Land and look at the other side of the coin. The Allies were confronted by gas warfare in April 1915: what did they have to counter the threat? In a nutshell, nothing. The dyestuff industries of Britain and France were practically non-existent, and their heavy chemical industry was not oriented to the production of complex organic substances on the scale of the German factories.

The first reaction to the Ypres cloud attack was to find every soldier with any knowledge of chemistry, give him a few minutes instruction, and send him to the front to organise and advise on the issue and use of the first protective masks. This was a temporary measure to try to counter the immediate threat while a more formal organisation was put together. Since, in the British army, there was no existing military department or appointee concerned with the questions arising from gas warfare, a Major Foulkes of the Royal Engineers was appointed 'Gas Adviser to General Headquarters' and given the entire task of assembling and training an offensive gas organisation. Four 'Special Companies' of Royal Engineers were formed and, since offensive action had to depend on what was available, they were trained in the operation of gas cylinders. By the time the war ended there were twenty-one companies forming the 'Special Brigade' and they were responsible for the use of cylinders, the 4-inch Stokes Trench Mortar and the Livens Projector.

To back up this effort a central laboratory was set up at British GHQ in order to identify new German gases and develop protective measures. At the same time medical research facilities were set up to investigate the problems posed in the treatment of the various types of gas casualty and to organise the medical services in the field so that they could implement these recommendations and be prepared for the new influx of casualties.

Eventually this dispersed effort was brought under central control by the establishment of a Directorate of Gas Services early in 1916 so that, fairly rapidly, the army had organised itself both to defend and attack in the chemical war. But the greater problem lay in the provision of the wherewithal to fight; without the specialised chemical industry, this meant setting up manifold agencies to perform the various functions which in Germany were all dealt with under the IG Farben structure.

A Chemical Sub-Committee of the Royal Society was rapidly put together, composed of some of the most eminent chemists and other scientists in the country -- Sir Oliver Lodge, Lord Rayleigh, Sir William Ramsay, for example -- while a similar 'Anti-Gas Committee' was formed to look into the protection side. Shortly after this the Ministry of Munitions was formed, and within this a 'French Warfare Department' was organised with, among other things, responsibility for gas warfare. A Scientific Advisory Committee sprang up and also a Commercial Advisory Committee where scientists and manufacturers could get together to hammer out the problems of research and manufacture.

For a further year there was constant change in this organisation as it was altered, tried, changed and argued over, and the committee, sub-committees and sub-sub-committees sprang up like weeds and flourished briefly until supplanted by a fresh re-organisation. It was control by committee at its very worst, with research gradually moving further and further away from manufacturing realities and the two of them out of touch with the actual users at the front. Eventually the whole arrangement was brought under central control by the setting up in October 1917 of a Chemical Warfare Department under Major-General Thuillier, who had formerly been Director of Gas Services in France.

The French had gone through the same sort of performance but they had got it out of their system more quickly. Beginning with a hastily-assembled Commission of soldiers and scientists in April 1915, in June a Chemical Warfare Research Committee was set up. In August this was superseded by three committees, one dealing with protection, one with offense, and one attending to liaison with the troops at the front. This worked reasonably well, but in July 1915 the whole lot were incorporated into M Albert Thomas's new Ministère de l'Artillerie et des Munitions under the control of a Directeur du Materiel Chimique de Guerre, and this effectively brought all the various aspects of chemical warfare under one unified control.

When America entered the war in 1917 they too had practically no chemical warfare capability. A small amount of research into war gases had begun late in 1916, though largely as a private venture, and although gas had by then been in constant use for two years in France the US War Department had made no provision for gas warfare at all. The first expedient was to divide responsibility between the

Major Foulkes, Gas adviser to General HQ', with Queen Mary

branches of the army which seemed appropriate: thus the Medical Corps was given the problem of protection, the Ordnance Corps the problem of procurement and supply of munitions, and the Corps of Engineers the task of raising, training and employing gas troops. As a planned system it suffered from the same defects as the original British and French systems, but it was made a good deal worse by the fact that the Americans were starting from scratch to enter a war in which the other contestants all had two years start on them. So that, for example, when the Medical Corps were peremptorily ordered to produce 1,100,000 respirators within twelve months, it had no facilities and no organisation with which to perform this feat. Similarly the Ordnance Corps had no facilities for producing gas and no private chemical companies were on hand to provide it.

In October 1917 a Chemical Service Section of the army was set up, but this, although producing a co-ordinating authority, could do little since it had no mandate to control research, production or gas warfare policy. Finally, in June 1918 the Chemical Warfare Service was authorised as a distinct branch of the army, and at last all the various activites were brought into one cohesive body.

But while the Allies slowly evolved

their organisations to wage war, there remained the task of actually producing the chemical agents, and, moreover, producing them in the vast quantities required to allow effective offensive operations. In Britain, for example, some twenty chemical plants were given contracts for various substances, but in the great majority of cases the products they were asked to make were completely new to them. They needed fresh plant, technical advice and extra staff, and all these had to be organised and produced by the various departments of the Ministry of Munitions. Eventually over seventy different establishments, the majority privately operated and the

American infantry adjust their masks during training

rest run by the ministry, were producing chemical munitions or their components, but the problems were enormous.

Perhaps the best illustration of the type of problem which had to be faced lies in the British production of mustard gas. When the Germans introduced this in 1917 they were fairly confident that neither Britain nor France would be able to manufacture it in quantity before the German army was victorious. Not knowing this opinion, the Allies resolved to adopt it themselves; the first problem was

simply how to make it. Mustard was a well enough known substance for it to be easily made in minute quantities in a laboratory, but producing it by the hundreds of tons was a different matter. Fortunately, analysis of mustard gas can tell the chemist how it was manufactured, and having found the German method steps were taken to organise a similar plant in England. The process was known as the 'Mayer' process, after the German chemist who, in 1886, devised it. The process begins with ethylene chlorohydrin, which is treated with sodium sulphate to convert it into thiodiglycol. This is then treated with gaseous hydrochloric acid to produce diethylene diglycol, or mustard gas. This sounds simple enough, but even the basic ethylene chlorohydrin is a difficult manufacturing problem, and the whole process is really only feasible to a well organised and well equipped organic chemical plant.

As it happened, a small pilot plant for the production of ethylene chlorohydrin had been built in England in 1915 in connection with the production of novocain, a local anaesthetic, and it was decided to enlarge this facility and use it as the basis of mustard gas production. But enlargement of a small pilot plant is a lot more formidable a task than simply drawing everything bigger, and while

US troops prepare for a gas attack

this was in progress a number of chemists looked again at mustard gas to decide if it might not be possible to make it some simpler way.

Professor Sir William Pope and his research team eventually, towards the end of 1917, developed a method of processing sulphur monochloride into mustard gas by blowing ethylene gas across it. Even this wasn't so simple as it sounds, but it was a good deal simpler than the Mayer process in terms of plant, and a factory was rapidly built by the Castner Kellner Company. The process was then improved by the Levinstein Company of Manchester, and the final product is usually referred to as 'Levinstein Mustard' to distinguish it from that produced by the Mayer process.

While this process was an improvement on the German method, it produced a mustard gas with a proportion of impurities, notably sulphur, and this was to cause trouble in the next stage, the charging of shell. As soon as the manufacturing process was settled, plans were drawn up to build a special charging factory at Chittening, near Bristol, to be operated by the Ministry of Munitions as 'National Filling Factory No 23'. Existing ammunition filling factories at Hereford and Banbury were also taken over and plans made to convert them from filling high explosive to charging mustard gas.

The contract for building the Chittening factory was given to a company who were familiar with the requirements of conventional ammunition filling plant, but beyond telling them that the material to be used in the new factory was a hazardous liquid, little was done to explain to them the exact sort of danger involved. The result is best described in the words of the engineer who had to run the place: "When the factory was taken over it was obvious that it had been built by engineers who, if they possessed any knowledge of mustard and its dangerous properties, did not succeed in applying it effectively to the construction. The whole installation gave the impression of having been designed for dealing with a liquid which could be handled with impunity...."

When the factory went into operation, the original charging machinery, which filled shell by gravity, was a constant source of trouble, and a lot of this stemmed from the sulphur impurities in the liquid gas. Another difficulty was the problem of convincing the work-people that it was, in fact, a dangerous substance.

"The majority of the operatives had no idea of the extremely dangerous character of mustard gas and appeared to think that the absence of strong smell and of immediate action was evidence of its innocent qualities. The result was that the gas was allowed to infect everything, and was even left lying in puddles on the ground. Innumerable buckets, bits of machinery, pieces of pipe, tomato cans (which were very popular for the purpose) and every kind and description of vessel which could, and frequently could not, hold a liquid, were to be found everywhere in the neighbourhood of the charging sheds, containing or at least contaminated with mustard. When, as frequently was the case, a certain amount of gas had to

Gas mask drill for US infantry

be disposed of, it was simply emptied into an ordinary two-gallon bucket and given to a labourer who could use his own discretion as to when and where it was to be emptied. On expostulating with one of these individuals, I was informed that the liquid was 'as harmless as mother's milk'."

As a result the casualty rate was never less than 28 per cent of the work force on each shift, and at one stage there was one casualty for every nine shells leaving the factory. Matters improved when a better design of filling machine, originating in France, was adopted, but this had barely got into operation before the armistice. In spite of all this, the factory managed to fill over 85,000 shells in four months of operation, though this was a long way from Brelch's plannec 25,000 a day. Had the factory beer completed (which was scheduled fo 18th November 1918) the final outpu forecast was 95,000 shells a week.

The French had more or less th same problems but managed to ge into production with mustard ga before Britain did. Their filling an charging plants had less trouble sinc they adopted the better design o charging machinery right from th start and never had the intense diffi culties that the gravity plant caused i Britain. Moreover they seem to hav had a greater sense of urgency; whil the British factories were operatin two shifts a day with civilian labour the French were running three shift round the clock and using militar

American troops pass through a gas filled section at the Battle of Chateau Thierry

labour, paying the soldiers a bonus of three francs a day extra for the hazardous work. One factory at Salaise, south of Lyons, had a planned ultimate capacity of 20,000 75mm, 2,500 105mm and 3,000 155mm shells per day. After a visit to a mustard gas manufacturing plant at Pont de Claix, a British expert reported:

"The factory works in three eight-hour shifts, one day's rest being given per week when possible.... There is no regular alternation of labour but after working for twenty days each worker is sent for five days to a rest house in the mountains. The works manager, who was due to leave on the following day, was quite ready for his rest cure, and this was even more emphatically true of the foreman who was superintending the filling of the drums.... The point was raised that the best method of avoiding casualties consists in a careful selection of the workers, who should be 'Très solide, très calme, bien réflêche'."

The significance of all this effort is seen in a very simple statement. The Germans fired mustard gas at the Allies on 12th July 1917. The Allies first managed to shoot mustard gas back on 9th June 1918 -- so the Germans had had eleven months of superiority with what was considered both then and in later years the best war gas ever developed.

Consummation

Construction of a German gas projectile position

With the introduction of mustard gas and the arsenical smokes the gas war entered into its final and most intensive phase. Though production on the Allied side was only just beginning to show results (Britain had managed to manufacture 8,000 tons of gas in 1916 and doubled it in 1917, though half of that production was chlorine) the questions of research and production were largely settled and the prime requirement was to decide on systems of employment which gave the best results.

Because their gas production programme and their geographical location favoured it, the British army became specialists in cloud attacks, and by the end of the war they probably knew more about this particular type of chemical warfare than did anyone else. The cloud attack operation as first devised by Professor Haber was a deceptively simple matter and doubtless looked most attractive when viewed from a laboratory in the Kaiser Wilhelm Institute: you decide the length of front to be attacked, calculate how much gas will be needed, divide that by the capacity of a cylinder to arrive at the number of cylinders needed, and then tell some soldiers to get on with it.

It is at this point that the rot begins to set in. Listen to an officer who had considerable experience of organising cloud attacks:

'Few people realise what hours of agonised effort were involved in the preparation and execution of a cloud gas attack... The "carrying-in" could not be spread over an indefinite period and usually took from two to six nights... Picture the amount of organisation and labour required to install 2,000 cylinders on a two-mile front. These cylinders would have to be assembled at a number of points in the rear of the line. No horse or lorry transport could assemble at such points before dark or be left standing there after dawn... In a normally difficult trench system, for a carry of a mile or a mile-and-a-half of communication trench, at least four men per cylinder are required to give the necessary margin for casualties and reliefs. This implies the organisation of more than 8,000 officers and men for the installation, with a fundamental condition that only small groups be assembled at any one point at any given time... to carry hundreds of pole-slung cylinders, already weighing as lead, round traverses on a dark night, is a feat requiring superhuman endurance. Therefore many "carries" finished with a hundred yards "over the top" through the parados wire to the near locality of the emplacement. This last carry was critical; a false step, the clatter of falling metal, meant drawing the fire of some curious and alert German machine-gunner… '

Nevertheless the British army persisted with the system since it developed a heavy concentration, though after the Livens Projector became standard the number of cloud attacks decreased. But research into cloud techniques continued in the hope of reducing the work, simplifying the organisation and doing away with some of the drawbacks. These drawbacks were firstly the fact that when the cylinders were opened the escaping gas made a loud hissing sound which was often heard by the troops in the enemy trenches, whereupon they had sufficient warning to put on their masks. Another was the sudden drop in temperature of the cylinders as the gas escaped, due to the absorption of heat from the surrounding air as the liquid in the cylinders changed into gaseous form. This caused frost and ice to form on the cylinders and valves -- sometimes blocking the valve and giving the cylinder operator a dangerous moment or two in clearing it, and, worst of all, it generally caused a mist to rise above the cylinders, another warning to an alert gas sentry on the other side. The final drawback was that the leading edge of the cloud was generally fairly dilute and thus, even if the other symptoms were not enough to alert the victims, they were still given a few seconds warning in which to take protective steps.

There seemed little chance of curing this last defect, but the others might prove amenable to mechanical research, and early in 1918 the British army developed what became known as the 'beam system'. Instead of bringing the cylinders right into the front line and spreading them for the full length of the intended cloud front, this system concentrated the cylinders into

Loading gas shells into the projector

72.

Cleaning the projectors after use

Projector emplacement constructed in a training programme

a small area well behind the lines. They were arranged on their sides in banks, four and five deep, on railway trucks and brought to the necessary point on the network of lines which proliferated behind the front; if necessary a fresh spur of track could be laid relatively easily. The cylinder valves were designed to be opened electrically and were of large bore so as to release the gas rapidly and rather less noisily, although noise was no longer particularly important. With several truckloads moved into place, the friendly troops in front were alerted to put respirators on and take cover and the electric circuits were energised. The valves all opened simultaneously and tons of gas were released to travel downwind in an ever-widening 'beam' until, if everybody had done their sums correctly, it hit the enemy trenches at exactly the correct width to cover the desired front.

The beam system worked exceedingly well; there was no advance warning to the enemy from days and nights of 'carrying-in', and nothing for the enemy aviator to see. It relieved hundreds of men from days of backbreaking and dangerous work. It was released too far from the enemy lines for the sound to give advance warning. And provided the wind was not too strong the cloud tended to cohere and roll forward on itself, so that the leading edge became as highly concentrated as the rest of the cloud.

The adoption of the beam system was so effective that it changed a lot of people's opinions on cloud gas operations. By the end of 1917 it had appeared that the technique had outlived its usefulness and that the balance between the amount of effort put in by the attackers and the amount of casualties suffered by the defenders was such that the cloud was no longer worth using. In June 1917 for ex-

Fuse-setting at a training school

Transportation of German projectors by a column of carriers

ample, the Germans had released 1,000 cylinders against the French and obtained only 375 casualties of whom but seven died. At Kovel the Russians released 4,000 cylinders -- 140 tons of mixed chlorine and phosgene -- to obtain 800 Austrian casualties of whom less than ten per cent died. But the beam attacks soon changed all that. Although only a few were mounted, they were highly effective. In July 1918 at Lens, the British amassed 5,110 cylinders carrying 148 tons of gas and accounted for 4,200 casualties.

On the face of it, there seems to be no good reason why, once the initial surprise attacks were over and troops provided with serviceable respirators, the cloud attack should succeed. But it did: and in a post-war analysis, General Foulkes, the mainspring of the British gas warfare effort, went so far as to say: 'given a favourable wind, every cloud gas attack is successful, more or less, even when it takes place in daylight and the troops opposed to it are provided with efficient respirators, have been warned of its imminence and actually see it coming... Every record we possess...proves this statement indisputably...'

The indisputable effect was largely due to good old human frailty: Foulkes cited an instance when, after a German cloud attack on the French under the worst technical conditions -- a bright moonlight night with a high wind -- 120 casualties were questioned as to why they had been caught. The answers were enlightening: 32 per cent admitted having been taken by surprise, 12 per cent removed their masks too soon, thinking the attack was over, and 52 per cent either put their masks on carelessly or allowed the fitting to go out of adjustment. The few remaining had faulty or inadequate masks.

In spite of this sort of opinion, and possibly because in the confusion of the mid-1918 battles the gas casualty

On the look-out for gas in a British observation post

74

figures were lost sight of, the Germans tended to regard the gas cloud as a minor problem and the Livens Projector as the most dangerous form of delivery. A British officer who was wounded and captured in May 1918 spent several weeks in a German hospital. Having been in the Special Brigade at one time, he was interested to see what the German experience of gas was, and he spent much time casually interviewing German officers and soldiers and gradually piecing together their opinions. He found that 'The Germans had no fear of cloud gas in any form and were surprised that we continued to use it' (though we should note that this was before the introduction of beam attacks) 'but the one form of gas they fear is that from the Livens Projector. They admitted serious casualties from them, and one

German troops lay out the lead pipes for a gas attack

A Livens projector in action

NCO who had been at Lens for some time said that they had made life there unbearable.'

With such a view commonly held, it is not surprising that the Germans rapidly undertook the development of a projector of their own. By late 1917 a number of pioneer battalions were equipped with an extemporised projector which fired an obsolete 180cm mortar shell. This was forced on them by the shortage of steel which prevented mass production of a more efficient form of projectile on the lines of the British Livens drum, and the gas content of the bomb was only about half that of the British projectile. In spite of this they used them to good effect, notably in the Battle of Caporetto on the Italian Front in October 1917, when they helped to inflict one of the worst defeats of the entire war.

They then sat down to produce a better projector than the Livens, and

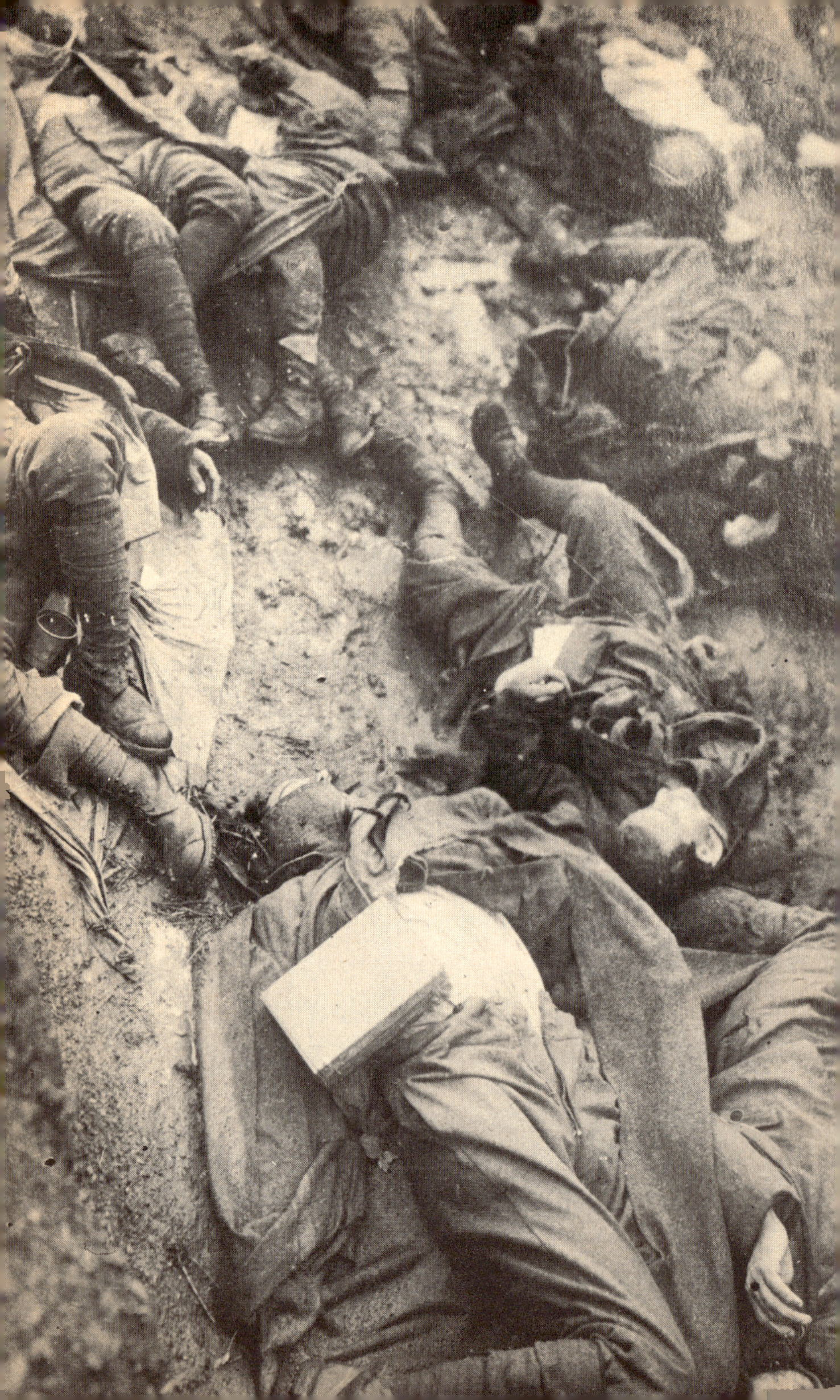

Lewis gunners in action in a front line trench during a gas attack

in the summer of 1918 began issuing a rifled model, firing a high-capacity shell and capable of reaching to 3,000 yards range. Great things were hoped for from this weapon, but in the event relatively few were made and they had little opportunity to use them before the war ended.

Apart from the physical problems entailed in emplacing it, the Livens Projector had one tactical drawback; it was the only missile-throwing weapon used with gas munitions which was absolutely useless for anything else. When it wasn't being used to throw gas it wasn't being used at all, and such a lack of flexibility has its drawbacks. Either you have a lot of expensively-trained specialist troops and their equipment sitting doing nothing when not engaged in shooting gas, or you send them off without their equipment to do something useful like mend roads or whitewash coal piles,

Italian casualties after the gas attack at the Battle of Caporetto

and then have the devil's own job to get them back and re-organised for the next gas task. For this reason the trench mortar was a highly regarded weapon on the Allied side; when not firing gas it could be laying smoke screens or firing high explosive shells for the infantry, which kept the mortar troops trained, occupied, and doing something useful all the time.

Although the Germans started off the trench mortar gas bomb business, they soon lost interest in it, largely because of the General Staff opinion that gas was not for harassment and attrition but for use in aid of definite operations. After the 90mm mortar was provided with a good bomb (to replace the two-glass-bottle model of the early days) the 17cm Minenwerfer was given a sizeable projectile known as the 'B-Mine' since it was filled with 'B-Stoff' (bromoacetone). This 97lb projectile carried 27lbs of gas and was

Flanders mud; unsuitable ground for high explosive shells

a most effective weapon. However, the principal German advocate of the trench mortar as a gas weapon was a Major Lothes, and after his death on the Verdun sector in 1916 the driving force was lost and interest flagged. Less than five per cent of the German gas output went into trench mortar bombs and, as a postwar report said, 'even those were mostly left lying about the countryside and not fired.' There seemed to be a possibility of a revival of interest in 1918 when a number of 17cm bombs were filled with mustard gas and fitted with time fuses in order to burst them in the air over the trenches and shower liquid gas onto the occupants. This was a good idea and it could have made life extremely trying for the Allied front line troops. But when the Germans tried it out, they were very unlucky; the wind blew the gas back into their own trenches. This happened twice, and that was that: the infantry commanders had problems enough trying to cope with enemy gas, let alone worrying about their own, and interest in the mortar lapsed once more.

The weapon which made most progress was, of course, the artillery shell, and its progress was not due solely to the mechanics of its construction. By 1918 that side of it had been fairly well ironed out, and efficient shells were in ample supply, though the fusing problem was never satisfactorily solved. Given that one could fire an efficient gas shell to where it was wanted, its efficiency thereafter depended on how it burst. In soft ground -- more often mud -- as was normal in Flanders, the normal fuses which were satisfactory with high explosive shells tended to react too slowly and allowed the gas shells to penetrate the earth before they burst. Because of this, a large

percentage of the filling was lost, either by being driven into the ground beneath or blown into the air by the cratering action of the half-buried shell. The ideal would have been a super-sensitive fuse to burst the shell on the surface or a time fuse to burst it above the ground, but although the British army developed a suitable combination of the two in 1917, the 'Time and Direct Action No 90', it was not found possible to put it into production without disrupting the manufacture of other and more vitally-needed fuses, and it never came into use.

The principal improvement in artillery gas warfare lay in the tactical handling and application rather than in the nuts and bolts. The German army undoubtedly led the field in this, largely because their better production facilities gave them the stocks of ammunition needed to put their ideas into effect. The development of artillery tactics as the war progressed is fairly well marked: originally the system was to fire a relatively intense but small area concentration whenever a suitable target presented itself -- troops forming up, reserves awaiting a chance to counter-attack, even a temporary traffic congestion at a cross-road or village. Early in 1917 the improvement in ammunition supplies and types allowed the growth of area bombardments, in which an enemy occupied area of quite large size would be engaged with as many guns as possible in order to get a heavy concentration of gas in the shortest time. Generally these bombardments were planned so as to place heavier concentrations on those parts of the area holding large bodies of troops, though the whole area would be laid under lethal attack. Another technique was to simply divide the area off on the map into 100 metre squares and allot a gun to each square in an attempt to get an even coverage.

Then, late in 1917, with the advent of mustard gas and the arsenical smokes, the German artillery entered a new phase. Complex instructions for the use of various combinations of gas in differing tactical situations were drawn up and the amount of gas ammunition produced reached astronomical figures. Much of the groundwork for this was done in operations on the Eastern Front where that notable artillery expert Lieutenant-Colonel Bruchmüller tried out various ideas until he was able to codify his findings into general firing instructions which were to become standard operating procedure. Bruchmüller was a great believer in the efficiency of gas shells; his first extensive use of them appears to have been against the Russian army at Witonitz in November 1916. Benzyl bromide (T-Stoff) and chlormethyl chloroformate (K-Stoff) were fired at individual battery areas with considerable effect on the efficiency of the Russian defensive artillery fire. In April 1917, when the bridgehead of Toboly on the Stochod

Lieutenant-Colonel Bruchmüller

was taken, Green Cross shell were used against both artillery and infantry, and in this battle Bruchmüller introduced a system of 'Gas Squares', areas containing enemy batteries which were to be inundated with gas as a whole, without trying to register precisely on the Russian battery in the square.

By now Bruchmüller had spread the doctrine of 'predicted fire' and practised his gunners in it. The usual artillery technique of that time was to fire one gun of each battery at its selected target until the range and bearing were found, and then to pass the data to the other guns so that they could begin effective fire. Obviously, against an alert enemy this only served to warn him of what was coming and this destroyed the surprise effect which was probably more important in gas shelling than in any other form of artillery fire. By using meteorologists to measure wind speed and direction, and by making allowances for air temperature and density, all of which affect the flight of the shell, it became possible to deduce firing data sufficiently accurately to allow concentrated fire to be brought down without the need for registering or ranging shots, and in this way the surprise element was brought back.

The first use of this technique by Bruchmüller seems to have been an attack on 19th July 1917 on the River Seret in Eastern Galicia. The counter-battery gas shelling began before dawn, to take advantage of the damp atmosphere to keep the gas concentrated, and the effect was a complete justification of Bruchmüller's theories.

When the Blue and Yellow Cross shells were introduced, Bruchmüller adopted them and allied them to his existing systems to develop the 'Buntkreuz' (coloured cross) system of mixing Green and Blue Cross in varying proportions with high explosive to provide an unheralded whirlwind of gas and explosions to whisk the German army across the Dvina river and into Riga. With 750 guns and 550 mortars, he arranged for counter-battery fire using 75 per cent Buntkreuz and 25 per cent explosive shells, while the infantry positions were shelled with 20 per cent gas and 80 per cent high explosive. This bombardment was concentrated into a five-hour storm of shells, and the subsequent attack was practically unopposed. Defensive artillery fire was negligible, many of the Russian batteries abandoning their guns owing to the intensity of the gas concentrations, and without doubt much of this was due to the success of Blue Cross gas, against which the Russian mask gave little protection. In all, 116,400 gas shells (388 shells a minute) were fired by the Germans to cause an estimated thousand casualties.

Shortly after this, Bruchmüller was

The gas cloud boils up from the point of impact of a gas shell

Above left: **A German ammunition column advances through wood contaminated by gas** ***Above:*** **Gas-masked Germans on the road to Soissons** ***Left:*** **A German transport in Eastern Galicia where Bruchmüller's technique of 'predicted fire' was first tried out**

moved to the Western Front, his talents being no longer needed on the Eastern with the Russian collapse. By now he was sufficiently well regarded for this theories to be listened to with some respect, and his hand is discernible in new standing orders for artillery which were issued in December 1917 and which laid down a series of rules for the employment of gas shell in different situations. Indeed, these instructions were so complicated and covered so many contingencies that they were really too involved to be a practical guide for the gun troops, but at least they gave a clear outline of what gas shelling could be expected to achieve if it were used properly.

The main feature of the new tactics,

as revealed in the German offensives of 1918, was a short surprise bombardment of terrific intensity, without any previous registration to warn the victims. The general plan was the same for each of the five big attacks, though, of course, the duration altered as necessary. About 100 guns per kilometre of front were allotted, and the first stage consisted of a ten minute bombardment of every target in the attack area with Blue Cross and high explosive shells in the ratio of 9 gas to 2 HE. After this came about an hour's intense bombardment of Allied artillery positions with mixed Blue and Green Cross and HE. The third phase, lasting about an hour-and-a-half, was a bombardment of infantry positions as well as gun areas, Green Cross shell being used only on targets more than 600 metres from the German line in order to prevent any danger of drifting back. Finally the attack began and the usual creeping barrage was fired with high explosive shells, while nominated batteries kept up a constant fire of Blue Cross shell 600 metres ahead of the advancing infantry; this distance was selected so that the gas would have dispersed by the time the infantry arrived.

Since the allocation of guns gave the Germans a superiority of about 5 to 1, it meant that an intense concentration could be dumped onto every Allied gun area very rapidly, and maintained for as long as the attack took to develop. The first of these attacks, on 21st March 1918, saw over two million gas shells fired, including large quantities of Yellow Cross fired on the flanks so as to isolate the attack objectives. The mustard alone accounted for 11,860 casualties and the Buntkreuz for another 3,000 and the attack was such a success that it finally converted a number of German infantry and artillery commanders, who still had been dubious about the effectiveness of chemical warfare, to

A gas shell strikes near barbed wire barricade

Gas in use in the German offensive of 1918

whole-hearted support. Demands for gas shell soared, and by the middle of the year 28 per cent of all artillery shells being manufactured were for charging with gas.

The Allied artillery effort never reached such heights of sophistication or volume, simply because there was neither the quantity of ammunition nor the variety of shell fittings available. The French managed, by prodigious efforts, to begin firing mustard gas in June 1918, and the British in September, and both these events came as an unpleasant surprise to the Germans. They were so convinced of their superiority in this field that when the first French mustard gas arrived they assumed that the French could only have acquired the gas by draining the contents from 'dud' German shells they had collected, a proposition so ridiculous that if it were not vouched for by official statements one would

German infantry push forward during the advance on Montdidier

Inspection of captured German gas projectors

have difficulty in believing it. For form's sake, it appears, their gas service analysed the French gas and were somewhat put out to find that it was not manufactured by the German process at all, and that the Allies had a production system in operation which was potentially capable of out-producing the Germans. As a result, plans were drawn up to build a new gas manufacturing plant to use the Allied method, but the Armistice came before it could be completed.

In the non-persistent gas field the Allies could not imitate the Buntkreuz technique; the only gases they had in artillery shell were the simple lachrimators and lung irritants, so that their artillery fire was more straightforward, though still effective for all that. But the German Blue Cross shell were closely examined with a view to copying them, and the idea discarded when it was realised that the detonation of the shell did not produce the substance in a fine enough form to do what was hoped for. General Foulkes later told the story of what happened next:

'In September 1917...one of my officers, Sisson, in a spirit of investigation put a pinch of DA (the British term for diphenyl chlorarsine, or Blue Cross 1) which had been extracted from a German shell on the hot plate of a stove in his room in my headquarters. The result was so remarkable that everyone was driven out of the house immediately, and it was found that the latest German mask, even when fitted with the extension which had been supplied to give protection against Blue Cross gas, gave no protection whatever against the cloud produced in this way.'

This was the start of one of the most secret Allied weapons of the war, the 'M-Device' or 'thermo-generator'. This was a canister about 8 inches in diameter containing an incendiary material and a charge of arsenical compound. When the incendiary material was ignited, the heat vapourised the chemical agent and drove off a cloud of particles which could penetrate any mask then known. The next problem was obvious; the Allies had to sit on the idea until they could perfect a mask which would protect their own troops, and with the difficult problem of devising a suitable generator and the even worse one of producing protection, the war was over before the 'M-Device' could be brought into use.

It was found after the war that the Germans had, in fact, made the same discovery, and in the summer of 1917 had produced a 'Gas Box' generator. The design was inefficient since the heat from the generator caused an up-draught of air, lifting the particulate cloud well above the ground where it was harmless, so a technique of discharging phosgene simultaneously was proposed. This, being a cold gas, would tend to cancel things out and hold the particulate cloud down to a dangerous level and also, of course, was the lethal back-up needed to take advantage of the smoke penetrating the respirator.

It appears that about 200 of these generators were taken to the Eastern Front and tried out in late 1917, but the test was a failure. Many of the boxes failed to ignite and most of the remainder malfunctioned in various ways. The best that could be said of the trial afterwards by one of Professor Haber's assistants was that it had been 'rein professorisch gemacht' -- done in the real professorial manner, which is pretty faint praise. Although the Germans appear to have appreciated that Clark smokes produced in this way were more efficient, they did little to follow it up. It seems that they were confident of the efficiency of Blue Cross shell, so confident that any other system was considered superfluous; there also seems to have been some high-level personal conflict which has never been satisfactorily explained and which helped to kill the 'Gas Box' idea.

A French and a British policeman on patrol in battle-scarred Bethune

Consolidation

An American field hospital for gas victims

With the end of the war came the inevitable accounting, and then the long drawn-out political manoeuvring of treaties and agreements, one object of which was to try and prevent a repetition of the whole sorry affair. So far as chemical warfare went, the first move was to include in the terms of the Peace Treaty a stipulation that 'The use of asphyxiating, poisonous or other gases and all analogous liquids, materials or devices being prohibited, their manufacture and importation is strictly forbidden' (Art. 171). Considering that every other weapon was laid under some form of restriction, this was unexceptionable. Another clause (Art. 172) demanded that within three months of the Treaty taking force 'The German Government will disclose... the nature and mode of manufacture of all explosives, toxic substances and other like chemical preparations used by them in the war or prepared by them for the purpose of being so used.' This was a little more demanding, but again, not surprising in the circumstances.

But when the full significance of the casualty lists began to make their

impression, and with the memory of pictures of blinded soldiers groping their way to the field dressing stations, and with the assistance of numerous individuals one met who suddenly went into paroxysms of coughing, smiled apologetically, tapped their chests and said 'Gas, you know', the outcry against chemical warfare began to increase. Discussions were long and involved on how to outlaw all chemical production, how to control such a ukase by inspection, how to set up an international body which could detect any attempt at research into poison gases -- in short all the stratagems which have become familiar in the last thirty years in the context of nuclear warfare. But eventually the truth had to be faced; that you could no more prevent a chemist discovering a war agent than you could prevent him discovering a cure for rheumatism or dandruff; even an inspector at the next bench could never be sure what was in the flasks and beakers.

So it came down to an attempt to reach some form of international agreement and the first attempt was at the Washington Conference of 1921-22. Ostensibly a conference on Naval Limitation, somehow a proposal for the universal prohibition of Chemical Warfare crept onto the agenda. This, after examination by a technical sub-committee, was rejected as being too complicated to be operable, and so a modified proposal was put forward. But it came to nothing; for it take effect, the five signatory nations -- The United States, Britain, France, Italy and Japan -- had all to ratify the agreement, and on consideration the French refused to do so. This had nothing to do with Chemical Warfare; doubtless had this been the sole question at issue, the French would have signed quite happily, but there were other clauses, notably a reference to submarines, included in the same convention, and the acceptance of any

Attending to gassed troops at a field hospital

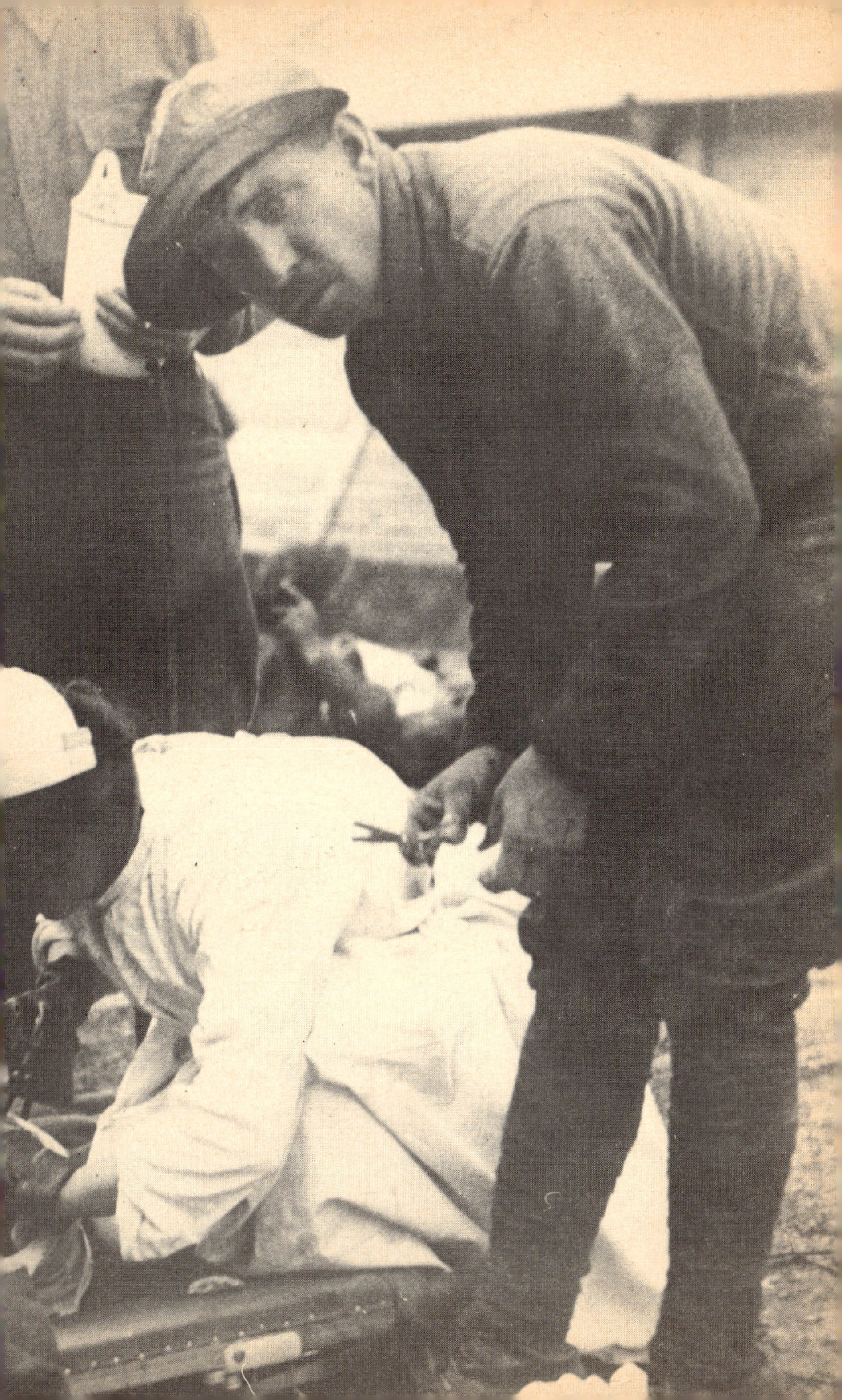

one item was not possible without the acceptance of all. The submarine item was the one the French jibbed at, but because of it the agreement on chemical warfare failed.

After this came the League of Nations, and though this august body had the gas question on the agenda for its entire life, the only concrete result was the Geneva Gas Protocol of 1925. In that year the League convened the International Conference on Supervision of the International Trade in Arms, Ammunition and Implements of War. The United States delegation put forward two proposals to prohibit the exportation of chemical munitions in order to 'lessen the horrors of war and ameliorate the sufferings of humanity incident thereto'. The delegates of the other countries, particularly the smaller nations without chemical industries of their own, were quick to point out the difficulty of identifying a chemical as being a component of a war gas when it could equally well be a component of an innocent industrial compound or even of a medicinal preparation.

As a result a move was made to include a clause in the proposed treaty on the banning of chemical warfare, but this idea was scotched when it was pointed out by the legal eagles that this was not within the Conference's terms of reference, and that delegates were not authorised to enter into any commitments or obligations on this subject. Eventually a legalistic way around this impasse was found by attaching the prohibitory clause as a 'Protocol' or appendix to the general agreement. This condemned the use of chemicals in war, included bacteriological warfare for good measure, and invited agreement to refrain from the use of such articles. Eventually over forty nations signed this Protocol; among the more important non-signatories were the USA and Japan.

It didn't end with that. For the next fourteen years the purple prose flowed freely as writers strove to outdo each other in painting the horrors of gas warfare. The remarkable thing is that nobody seems to have written a book about the horrors of, say, the .45 bullet and the 9.2 inch shell. It was, apparently, quite acceptable, even admirable, to have the back of your head taken out by a lump of lead or be disembowelled by a flying shard of steel, but entirely reprehensible to be made to sneeze or vomit for an hour.

The fact of the matter was that, certainly as things stood until 1936, gas warfare was not the inhumane activity which the propagandists painted it; it was, on the contrary, the most humane system of warfare possible -- always assuming you can allot degrees of humanity to warfare. Anyone who doubts this is invited to consider one

simple statement of fact: that the number of deaths due to gas during the First World War represents one-third of one per cent of the total casualties. Even taking the grand total of gas casualties, both killed and wounded, the total still adds up to less than five per cent of the total butcher's bill.

Moreover the after-effects were a good deal less than was generally believed; in 1937 figures published in the United States on American gas casualties who had been under observation since 1918 showed that, for example, pulmonary tuberculosis from the effects of gas were 2.45 cases per 1,000 men gassed, while the normal rate of occurrence of pulmonary TB among peacetime soldiers was 4.3 per thousand; and of 812 US soldiers blinded during the war, only 33 -- or 4 per cent -- were blinded by gas. British experience was similar; General Foulkes pointed out that of 4,575 cases checked, which had been serious enough for the wounded man to be evacuated from France to a hospital in England as a result of being gassed, 94.5 per cent of them were fit within nine weeks. And examination of Pension Board records in 1920 showed that in less than two years from the end of the war only 12 per cent of those who had been gas casualties

American crew prepare to decontaminate a shell hole

Doctor W Lee Lewis, inventor of Lewisite

were sufficiently disabled to warrant a pension. Moreover, of all the disability pensioners of the British armed forces after the war, only two per cent were disabled by gas.

In view of this, one is entitled to ask the obvious question; if gas did so little was it worth using it at all? The answer to that is certainly in the affirmative; the table shows that the total gas casualties were 1,286,853, and any weapon which puts a million-and-a-quarter men out of action is a weapon worth using. Moreover the amount of effort required to produce this million-and-a-quarter casualties was proportionately less than the amount needed to produce the other 95 per cent making up the total. It has been estimated that of the non-gas casualties, half were produced by artillery fire using high explosive and shrapnel shells; it has further been deduced that 1,389 million such shells were fired during the war, which is roughly 110 shells per casualty. On the other hand 66 million gas shells were fired to produce 75 per cent of the recorded gas casualties (the remainder being due to cloud attacks) giving only 68 shells fired per casualty, so that gas shelling was almost twice as effective as the more conventional ammunition in producing casualties.

But these are merely mathematical exercises; figures, notoriously, can be adduced to prove anything. But the inescapable fact is that gas caused a million-and-a-quarter casualties of whom 93 per cent lived to talk about it afterwards, some 12 per cent of them suffering some degree of permanent disability. And while it might be a terrible thing to have to live the rest of one's life with damaged lungs or impaired sight, it is a far worse thing to live the rest of one's life as a legless or mutilated cripple.

These figures, and others like them, were not widely broadcast in the 1920s though; gas was unpopular, and the whole question was treated very gently, on the principle of 'take no notice and it might go away'. But the military personnel of the world knew about the figures, and they were determined that should the situation arise again, their chemical service would not be found lacking.

This, of course, was a very delicate matter indeed; while the soldiers might see it that way, the governments very often didn't; the American army, for example, was strictly forbidden to do any research into toxic gases for several years. Anti-war feeling was high, and this, increased by the anti-gas revulsion, meant that any military man who as much as suggested gas warfare in public was half-way to the pillory before he closed his mouth. In order to avoid the accusation of war-mongering, it became necessary to employ various subtleties; no-one who condemned gas warfare could possibly object to research into defensive measures with a view to protecting the nation from gas attack by an aggressor. Given this, it was a simple step to point out that it was therefore necessary to study gas warfare's offensive aspects in order to be able to calculate what sorts of defence would be necessary. So every nation prefaced its military texts with some statement of this nature -- this one from a military training manual of 1922: 'Methods of defence must therefore be studied and developed and this can only be done efficiently if methods of offence are investigated at the same time. It follows therefore that nations must continue to study and develop chemical warfare.'

The study and development which took place was largely a continuation of work which had been in progress when the war ended. Both Britain and the United States had built up large test and experimental stations, the British at Porton on Salisbury Plain and the Americans at Edgewood Arsenal, and these establishments were retained after the war and set to the task of improving what was already known and developing new materials for both attack and defence. A mission was sent to Germany in February 1919 to obtain the information demanded in the Peace Treaty, and further information was acquired in 1921-22, until the secrets of the German gas effort had all been obtained. An interesting and slightly prophetic remark is recorded in one report: a German officer, obviously tired of being pestered, was driven to remark 'Why do you keep asking questions about mustard gas? You know very well that we will use something else next time!'

The allies certainly intended to use something else; in 1917 a Dr W Lee Lewis produced chlorvinyl dichlorarsine in an effort to produce a substance which combined vesicant action with systemic poisoning, and this became known as Lewisite. Mustard gas had its drawbacks, the principal one being that its action was very delayed, a secondary one being its tactical limitations due to its extreme persistence. On the other hand the systemic arsenic poisons developed by the Germans, the Blue Cross gases, had proved disappointing in use though on paper they had promised well. The Germans had appreciated these points and in 1918 had produced ethyldichlorarsine -- known by them as 'Dick' -- which was a lung injurant as toxic as phosgene, a powerful irritant, and also had a vesicant action, though not so severe as that of mustard. Finally, and most important in German eyes, it was non-persistent. Unfortunately for them, the development came too late for it to be used to any extent, and reliable reports on its combat effectiveness are non-existent.

Similar reasoning produced Lewisite, but owing to difficulties experienced in translating the laboratory process into successful quantity production it was not available in time to be used; the first batch was made in the United States and was ready for shipment in November 1918, and on the signing of the Armistice it was taken out and dumped at sea since its keeping qualities were an unknown factor. Postwar investigation revealed German researchers had also found the compound but they had rejected it as being insufficiently lethal, which shows how wrong you can sometimes be.

Lewisite was a persistent gas which took effect much faster than did mustard, and moreover the arsenic content could be absorbed through the skin. It was fairly easily destroyed by hydrolysis in moist air and when sprayed on watery ground, but one of the decomposition products was another toxic vesicant agent which was water-resistant, so that in effect Lewisite was a highly persistent material. All this, of course, led to wild speculation when, bit by bit, it became public. I can well remember in the middle 1930s that it was generally called 'Dew of Death' by some of the more sensational periodicals, and it was solemnly promised that one drop of this stuff on the skin was sufficient to cause death in seconds. In fact, it was nothing quite so drastic, but nevertheless it was an extremely lethal agent and its burning effect on the skin was a good deal faster than that of mustard. As to the systemic poisoning effect through the skin, the official rating was that the average man would be killed by the application of thirty drops to his skin; though how long the drops had to stay there to take effect is something which the textbooks all seem to ignore.

Lewisite became fairly well known during the inter-war years, largely, it seems, because its existence was announced in a fit of post-war exuberance, doubtless to reassure people that the Allies were as capable of devilish invention as the Germans and could have mounted a satisfactory gas capability had the promised 1919 offensive been necessary. But after that, the curtain descended; more than any other armament field, chemical warfare became the subject of rigid security, and very little hard fact on the development of chemical agents has emerged since. Many of the lines of research which were being opened up in 1918 were continued, but the results and the reasoning rarely became public knowledge. To give one example; a report on German research written in 1919 by an Allied investigator contained the sentence 'It is interesting to note...that sesqui mustard was being investigated, though there is no evidence to show that the Germans had any intention of using it as a war gas.' That was the last ever seen in print about sesqui mustard gas; Dr Sartori, the distinguished Italian chemist and expert on war gases, published a most comprehensive work in 1939 in which scores of compounds were described and discussed, but there is no mention of the mysterious 'sesqui mustard'. Another minor mystery is Arsine gas; this was also investigated during the war and dismissed as being useless,

The Scots Guards wear their latest gas masks, 1927

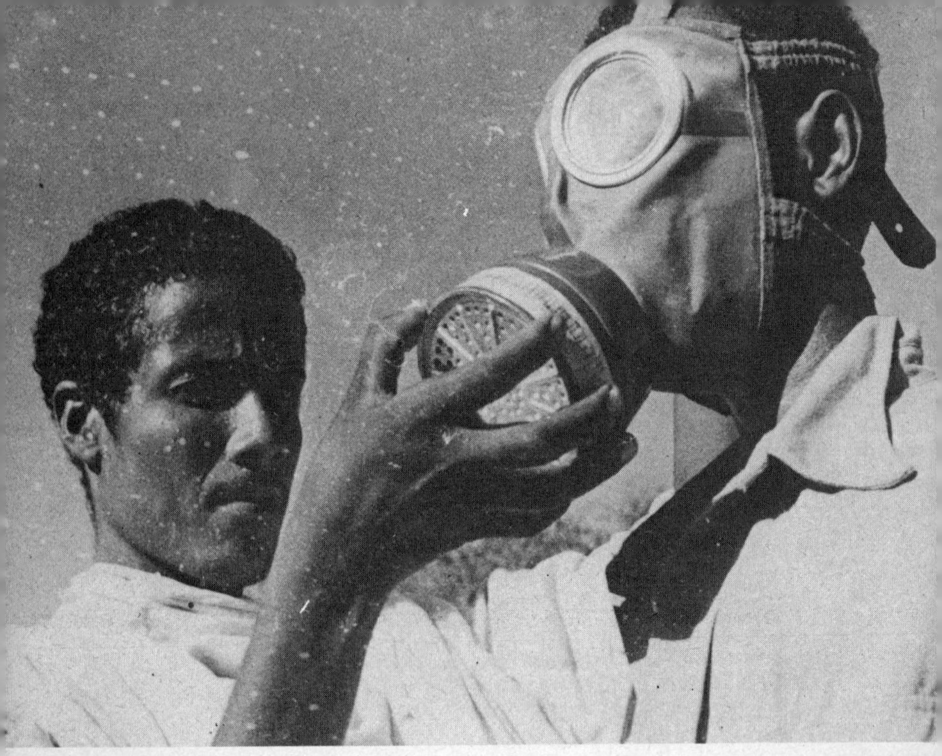

Gas masks in use by the Ethiopian army

the Germans doubting whether it would be possible to achieve a worthwhile concentration owing to various physical features of the gas. Yet in 1935 the official British gas defence handbooks all listed Arsine as a likely war gas; but without any explanation of how and why opinion had changed over the years.

Another late war development which reached fruition afterwards and which was publicly announced was 'Adamsite' or 'DM', diphenylamine chloroarsine, christened by the Americans after its discoverer Major Roger Adams. This came about in an American attempt to produce diphenylchloroarsine -- 'DA' -- the German Blue Cross Gas. In efforts to produce the gas by the process used in Germany considerable difficulties were encountered, which led to a decision to seek a more simple method. During this study, both British and American chemists found that DM, different but closely related to DA, could easily be made in large quantities. This became one of the principal developments in the peace years, as did a refinement of mustard gas known as 'nitrogen mustard'. At the same time there was much investigation of gases which had been summarily rejected in wartime since they posed problems in connection with manufacture or use which were felt to be too difficult of solution in a short time. Now that time was available the long list of 'possibles' was brought out and reconsidered.

The other field of research and speculation was in the systems of delivery or use of the war gases; designs of artillery shells, mortar bombs and Livens drums were improved, but the greatest question mark hovered over the airplane. The possibility of using gas in aircraft bombs had been considered during

An open-air lecture demonstration on the use of gas masks for infants in Paris, 1939

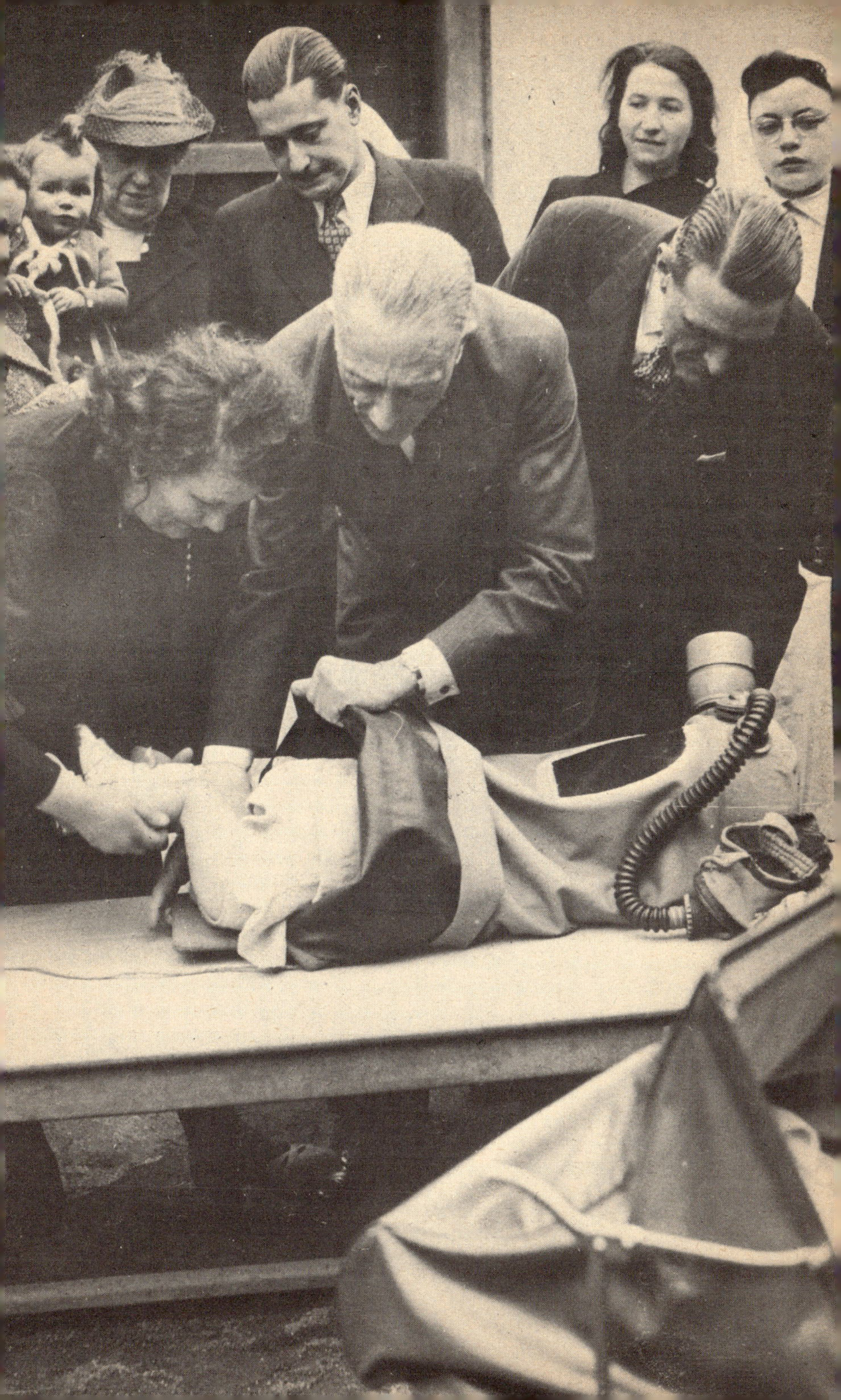

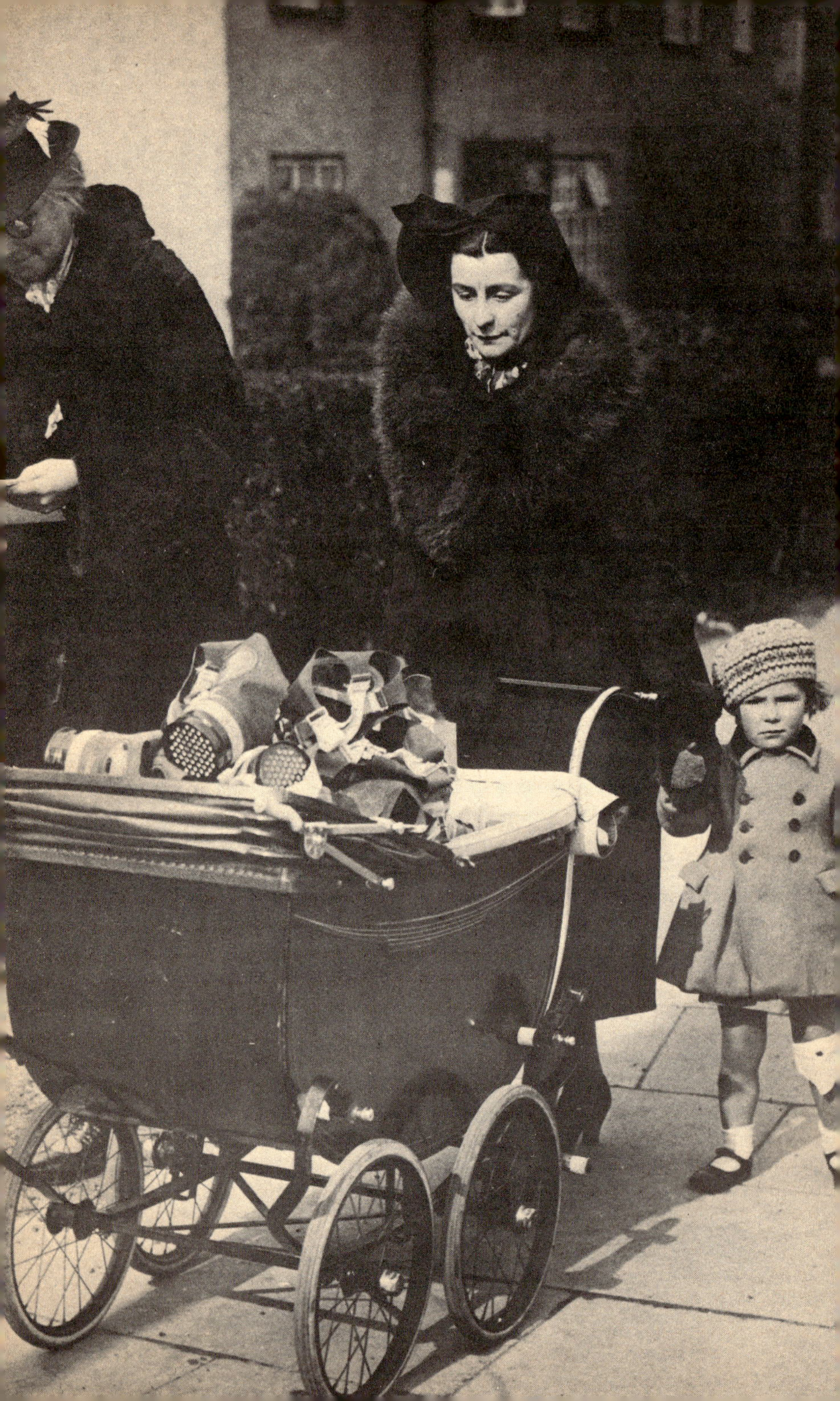

The possibility of chemical warfare is guarded against; lady visitors are issued with gas masks at Chatham

the war; on the Allied side it was largely theoretical, since there was insufficient gas production to satisfy the demands of the ground troops. The Germans had gone so far as to fill, by the end of the war, eight bombs with phosgene ready for testing, but the General Staff would not agree to the use of aircraft gas bombs since they considered them only of use against large targets -- military encampments or towns -- and they were afraid of reprisal raids against the Rhine towns if they were to initiate this form of warfare.

In postwar years the connection between the airplane and poison gas was so staringly obvious that it became the principal discussion point. The bomb was the first system of delivery which came to mind, but this was soon followed by proposals for spraying liquid gas from airplanes at low altitude, this being thought a highly effective method for distributing vesicant gas against troops in the field.

Delivery of gas masks by perambulator at Sidcup

Generally, in the middle 1930s, the gas problem seemed to die out of public discussion, but it was suddenly revived in 1935 with the revelation that the Italian Air Force had used gas against the Ethiopians, in spite of the fact that both the combatants in the war had signed the Geneva Protocol. With this, and with unfounded rumours of the use of gas in the Spanish Civil War and the threatening situation in Europe, the dangers of gas warfare suddenly became an important issue and steps began to be taken to protect the civil populations of various countries against the possibility of gas attacks by aircraft. Gas-proof shelters, the issue of respirators to the civil population, air raid drills, the formation of air raid precautionary services, all these were put in hand

PRIVATE
WARNING
THESE
SLOPES ARE
DANGEROUS

A rescue party makes its way through a cloud of gas during a dummy attack

in 1937-38 in every major country who considered the threat to be viable. It has been suggested that much of this well-publicised activity was designed to inform would-be aggressors that, since the country in question was alert to the problem and had precautions well in hand, such air attacks would merely be a waste of effort. This is perhaps an extreme simplification; there was undoubtedly a considerable fear of gas attacks at that time. I can still remember that in 1938, as a school-boy, I learned a little jingle which acted as a mnemonic to remind us of the identifying features of the various war gases; 'Phosgene smells like musty hay' ran one line. The trouble was that neither I nor any of my contemporaries had the slightest idea of what musty hay smelt like. Other gases, we were told, smelt of everything from Geraniums (Lewisite) to garlic (Mustard) or even pear drops (Bromobenzylcyanide). Any unusual smell in the street and my hair stood on end.

But gassing the population of a city was not quite the cut-and-dried exercise which the sensational journals promised. A single squadron, we were assured, could lay a city to waste in minutes. But the hard facts were against it; the aerial bomb of 1938 was, at most, a 300-pounder, and it was estimated that such a bomb would put down a lethal concentration over a 200 yard diameter circle. At that rate it would have required some 12,000 bombs to be dropped simultaneously to deal with London. Moreover they would only have produced a lethal concentration at ground floor level, and even then it would have dispersed within ten minutes. And this, of course, presumed still air and various other ideal conditions unlikely to be fulfilled in practice.

No matter; the population of Europe was apprehensive. The gas mask became a normal item of decoration on the hat-stand, air raid wardens checked the gas-proofing of their shelters, and odd-looking plates, covered in a sickly yellow paint, which, we were assured, would turn red if anyone as much as mentioned gas within earshot, sprang up all over. Europe was prepared.

The chemical war goes on, as part of the vast programme of preparedness; no one knows what the Soviets have up their sleeves, but since they began from the same captured data as did the Allies in 1945, it would be naive to suppose that their present situation is any great distance removed from that in Britain or the United States. Moreover, recent years have seen the arrival on the scene of some very peculiar substances which, without being lethal, could still be highly effective war agents. LSD, Lysergic Acid Diethylamide, has achieved considerable notoriety as an hallucinatory drug; the same sort of effects can be achieved by agents disseminated as gases or aerosols. Instead of killing one's enemy, and then having the problem of tidying up afterwards, a dose of some hallucinogen could 'send him on a trip'; render him incapable of defending

himself or acting in a rational manner until he came to his senses in a prisoner-of-war cage. Similar agents can induce a feeling of euphoria which overcomes motivation and training, so that defenders greet their attackers with smiles and open arms: other agents simply induce a fatalistic lassitude in which the victims sit passively, not caring, while the invasion rolls past them. The possibilities are endless -- and, I might say, pure conjecture, since, again, little has been made public on this score.

Twenty-odd years ago I was talking to a retired soldier who said that he had been concerned with chemical warfare. 'Gas?' I said. 'But that's all finished now, surely? I mean, we've got the atomic bomb, and all these missiles they're talking about...'

'Don't you kid yourself,' he smiled. 'Chemicals can do things that will make your atom bomb look like a child's firework.'

And that was twenty-odd years ago.

Gas drill for London policemen; a masked decontamination squad during training

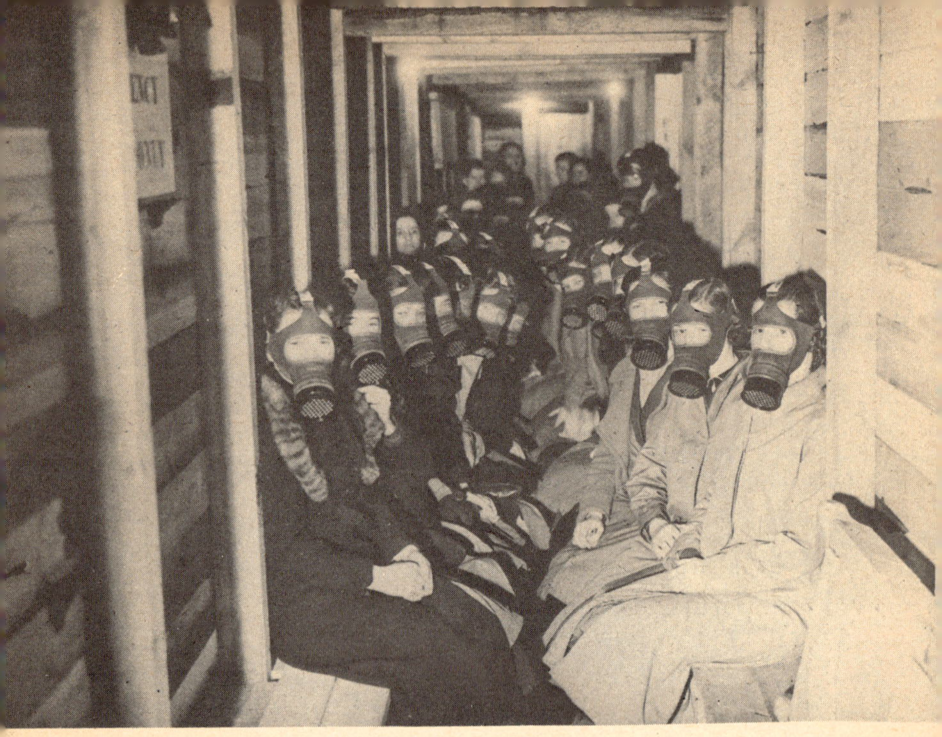

Above: Gas-masked girl employees in an ARP shelter. *Right:* A demonstrator using the latest apparatus for fighting gas. *Below:* A gas-proof baby carriage invented by a Kentish nursery man

Prognostication

The Chinese assumed that they would meet gas in their war with Japan

When war broke out in 1939, Great Britain immediately sought assurance from all the countries involved, or even likely to be involved, that they would abide by the provisions of the Geneva Protocol and abstain from the use of gas. All the countries questioned replied that the Protocol would be observed, even Japan, which was not a signatory and which had been under some suspicion of using gas in the campaigns in China and Manchuria. Nevertheless, knowing that such assurances might well be forgotten in the future, Britain took steps to build up a gas warfare potential. Productioı of chemicals was organised, charging factories set up, and the production of stockpiles of gas and artillery shells put in hand. The Livens Projector, which had been more or less ignored for years, was revived and production of drums and fuses began. Spray devices for aircraft, together with bulk gas for them, were procured and placed in store for the Royal Air Force. Research continued into both offensive and defensive aspects of chemical warfare, and at least one new weapon was introduced; in the besieged conditions of Britain in 1940, when every possible anti-invasion measure was examined, the 'Squirt' was devised.

The 'Squirt' was simply a pressurised cylinder of hydrocyanic acid on a man-pack frame, with a short hose and trigger-operated nozzle, and it was intended as a last-ditch anti-tank device. The operator concealed himself in some convenient spot and awaited the arrival of an enemy tank; he then ran forward and pressed the trigger to deliver a stream of prussic acid at the air intakes of the tank, in the hope that a lethal concentration of fumes would build up inside and dispose of the crew. Doubtless it had all been worked out in theory, but since the invasion never came, it was never put to practical test.

The Germans had also devised a chemical warfare adjunct to attacking tanks, but although it was used it seems to have had no noticeable effect -- rather like the shrapnel shells at Neuve Chapelle. At the outbreak of war the standard German infantry anti-tank weapon was the PzB 38 anti-tank rifle, firing a 7.92mm bullet from an over-sized cartridge. The bullet carried a hard steel core, to give it the necessary penetrative ability, a tracer element so that the firer could see where it went, and a tiny capsule containing 0.4 of a grain of lachrimatory gas. It all sounds very much as if it were another theoretical idea put into use without much in the way of

Gas-masked Japanese troops attack during the Battle of Shanghai

practical testing. The purpose of the gas capsule was to incapacitate the tank crew after the penetration of the bullet, but nobody appears to have noticed its effect, and its presence was not even suspected until some specimens of ammunition were captured and dissected.

At the time, of course, nobody could guess how the other side viewed gas warfare, and as a result everybody prepared for the worst. Britain, the United States, and Germany all provided their troops with respirators and anti-gas clothing and equipment and drilled them in their use. Retaliatory capacity was available, but in truth it was relatively small: it has been estimated that in 1944 the Allies could have engaged in all-out gas warfare for no more than a month, by which time the ready stocks would have been used up and the effort would have been brought to a standstill while fresh supplies were built up. In default of

knowledge, the intelligence agencies were put on the trail, but even they came up with very little solid information; there were rumours of new German gases of unparalleled lethality, but these were discounted as being the usual sort of rumours which fly round in wartime. Besides, the same agents were diligently spreading the same sort of rumours about Allied gases to try and put the wind up the Germans.

Once the tide of war moved in the Allies' favour and the invasion of Europe was planned, the question arose of whether the Axis powers might not fall back on gas as a defensive weapon in order to deny the Allies either a foothold in Europe or at least entrance into Germany. General Bradley afterwards said that had the Germans used a persistent gas on Omaha Beach, the effect might well have been decisive. Consequently a stockpile of gas munitions was built up in Britain and other areas, to be used in retaliation should the need arise. And from this came the war's only gas casualties.

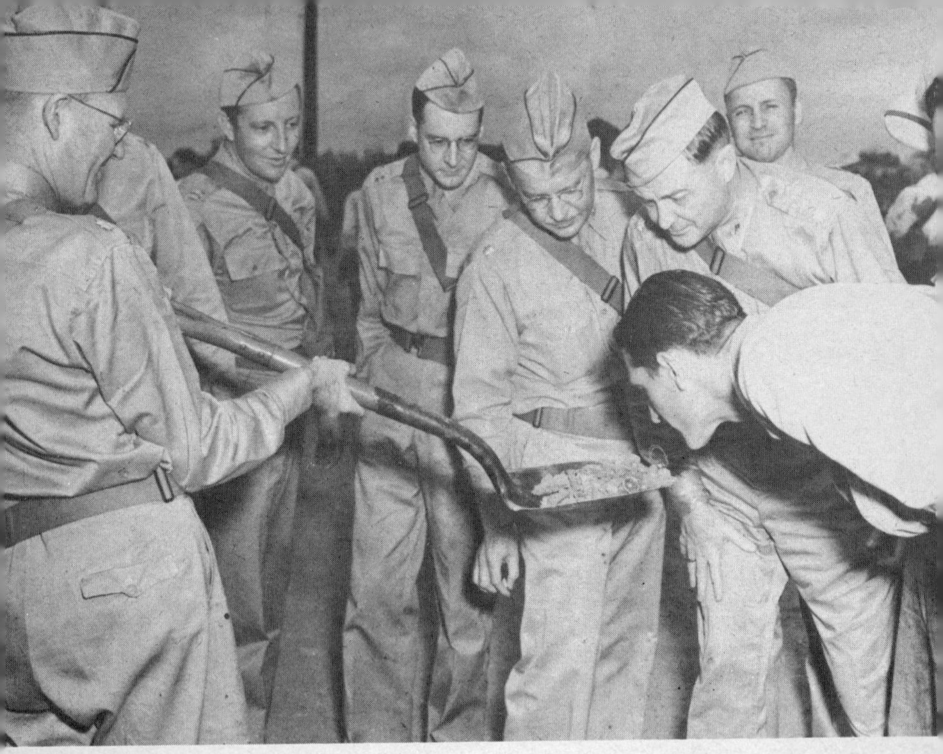

American officers become acquainted with mustard gas at a lecture during the Second World War

In December 1943 the SS John Harvey, a freighter, docked in Bari, Italy, part of a convoy of munitions and stores for the Allied armies in Italy. On board, among other types of ammunition, was a consignment of artillery shells charged with mustard gas. A sudden air raid on the freshly arrived convoy caused considerable damage, and many ships, including the John Harvey, were set on fire. The presence of the gas-charged shell was not known; and when the John Harvey's cargo exploded, the area surrounding the ship was immediately contaminated with several tons of mustard gas. As a result eighty-three men, principally sailors who had been swimming for their lives in mustard-contaminated water, died within a few days, and over 500 more were hospitalised. Had it been appreciated that mustard gas was the culprit, remedial measures might have been taken in time to save more lives, but the security was so tight that scarcely anyone knew of it until it was too late.

Apart from this unfortunate affair, there was no gas casualty list as a result of the Second World War. The reasons are fairly simple; in the first place, no nation was willing to be the first to use gas, partly because of the stigma of being the first to break the Protocol, and partly because there was always the fear that the other side might well reveal a considerable superiority. This latter was a constant factor; the Germans, for example, basing their assumptions on the quality and lavishness of Allied anti-gas equipment, assumed the Allies to be capable of a much greater offensive effort than they in fact were.

A second reason for the absence of gas warfare was that few soldiers either wanted it or had much faith in it. In spite of the records of the First World War, and in spite of the fact that many of the higher commanders had fought in that war, gas was largely an unknown quantity to the majority of the higher echelons. In the British army it was well appreciated, but it was considered solely as a weapon for defensive use or for use in static situations, and its use in fluid warfare was not thought to be practical. The Americans were averse to using gas at all, placing little reliance on its ability to be decisive, although there is evidence that its employment was contemplated when the invasion of Japan was under review; but only contemplated -- no decision was ever reached. And in Germany the prime factor, believe it or not, was Adolf Hitler's personal abhorrence of gas as a war weapon, stemming from his own experience of it during the First World War. Moreover the German General Staff were averse to using gas because of the possible effects on international opinion. Yet of all the nations at war between 1939 and 1945, the Germans were in the best position of all: the rumours had, for once, been right. The Germans did possess a super-gas.

In 1936 a Doktor Gebhardt Schräder was conducting research into organic phosphorus compounds for the development of insecticides and weed-killers for agricultural use. One of his discoveries was ethyl-dimethyl-amido-phosphoro-cyanidate; it proved to be a highly efficient insecticide, and it promised to be an equally efficient war gas. Further development followed and eventually it was decided to put it into production as 'Tabun', and a special factory was built near Dyhernfurth on the River Oder. The engineering problems of converting the laboratory reaction into a production technique were formidable, but by 1942 the factory was in operation with a planned eventual production figure of 1,000 tons of Tabun per month. Some 15,000 tons are believed to have been produced before the advance of the Soviet Army over-ran the factory, since which time nothing has ever been said about the place; it seems likely it was dismantled and taken back into Soviet Russia.

Since the organic phosphorus field appeared to hold promise more research effort was put on the trail, and in 1938 another composition was adopted for military use. This was isopropyl methyl-phosphoro-fluoridate, discovered by the same team who had perfected Tabun and now named Sarin. Plans for production were drawn up, but the manufacture of Sarin turned out to be even more difficult than that of Tabun, and although a pilot plant had been built and operated before the war ended, no production for service use ever took place.

Lastly, in 1944, 'Soman' was discovered; this was pinacolyl methyl-phosphoro-fluoridate, considerably more lethal than the other two but still only in the laboratory stage when the war ended.

Even with these three agents discovered and one of them in production, the Germans were still apprehensive of Allied capabilities, since they were under the impression that these or similar compounds had been discovered in Britain. The basis of this belief was that, for totally different reasons connected with other developments, the British censors had asked scientific periodicals not to publish research papers or speculative articles on various subjects, one of which was the organic phosphates. This sudden absence was noted by Germany, who immediately assumed that it was because of the war gas application, and from this deduced that the secret was not their sole property.

The Allies, though, had no inkling of these substances, and thus when, towards the closing days of the war, a dump of chemical munitions was found, there was a certain amount of curiosity about what the shells and bombs actually contained. The marking was a single green ring and the letters 'GA', which meant nothing that any allied agency had ever heard of. The contents were examined by some unknown hero who reported

US officers wearing gas masks during field exercises of the Chemical Warfare Sectio

'The chemical filling is dark brown with a sweetish smell of chlorobenzene...it is a new type of filling containing 20 per cent chlorobenzene and an arsenic derivative which is under investigation. It is not vesicant but somewhat poisonous by penetration through the skin...and is relatively persistent.'

Eventually about half-a-million shells and over 100,000 aircraft bombs charged with this substance were found in dumps scattered throughout Germany, although, according to one source, the aircraft bombs were not an operational stock but had been filled in order to use up the stock of gas at Dyhernfurth before the factory was abandoned to the advancing Russians. After analysis of the mysterious liquid, which was of course Tabun, the Allied chemists were astonished; when they later discovered Sarin and Soman and examined their properties, they were aghast. These were agents up to thirty times more lethal than anything known before, and, moreover, agents against which there seems little defence -- which may well have been one of the reasons for German reluctance to use them. These were the nerve gases.

The nerve gases, or 'G Agents' as they became known from the German terminology of GA, GB etcetera, operate on the human body in a manner quite different from previous war gases. The functioning of the body depends entirely on the nervous system transmitting impulses, and these impulses lead to the formation of a compound called acetylcholine at the nerve junctions. This compound is neutralised by an enzyme known as cholinesterase, which by its action on the acetylcholine 'cancels' the impulse on the nerve path and leaves the system open to receive a fresh signal. The G Agent attacks the cholesterine, converting it into a different compound, one which does not have the cancelling effect on the acetylcholine. As a result the nerve junctions remain blocked and fresh impulses cannot pass; the body's functions are no longer under the brain's control. The effects run from a running nose, through difficult breathing, distortion of vision, nausea, vomiting, involuntary micturition and defecation to convulsions, coma and eventual death.

This catalogue of symptoms is produced either by inhaling the fumes or by contact with the liquid, and the course of events is extremely rapid. In a lethal concentration, Soman can cause death within one or two minutes, while the others are only slightly less rapid.

As may be imagined, no sooner did the Allies appreciate the significance of what they had found than an impermeable security curtain fell on the subject, and, as far as the man in the street is aware, we are today in the same place as we were in 1945, since there has been very little ever made public about chemical warfare development since then. As these words are being written, an announcement in the press was made that the United States is to destroy large stocks of nerve gas owing to their becoming redundant; a new type has been adopted. This, we were told, is a 'binary' nerve gas, two liquids each harmless in itself which combine to form a potent lethal agent. They are to be filled into special artillery shells in two containers, so that the liquids will never meet until the shell is detonated at the target, allowing the two containers to break and the substances to mix. This will doubtless be a consolation to many people in the United States who have been worried about the possibility of an accident releasing stocks of nerve gas onto an unsuspecting population, a fear which was reinforced by the unfortunate affair some years ago in Utah, when a test cloud moved in the wrong direction and annihilated a flock of sheep.

A young militant, masked against tear gas in Ulster

IRELAND